Sustainable Civil Infrastructures

About this Series

Sustainable Infrastructure impacts our well-being and day-to-day lives. The infrastructures we are building today will shape our lives tomorrow. The complex and diverse nature of the impacts due to weather extremes on transportation and civil infrastructures can be seen in our roadways, bridges, and buildings. Extreme summer temperatures, droughts, flash floods, and rising numbers of freeze-thaw cycles pose challenges for civil infrastructure and can endanger public safety. We constantly hear how civil infrastructures need constant attention, preservation, and upgrading. Such improvements and developments would obviously benefit from our desired book series that provide sustainable engineering materials and designs. The economic impact is huge and much research has been conducted worldwide. The future holds many opportunities, not only for researchers in a given country, but also for the worldwide field engineers who apply and implement these technologies. We believe that no approach can succeed if it does not unite the efforts of various engineering disciplines from all over the world under one umbrella to offer a beacon of modern solutions to the global infrastructure. Experts from the various engineering disciplines around the globe will participate in this series, including: Geotechnical, Geological, Geoscience, Petroleum, Structural, Transportation, Bridge, Infrastructure, Energy, Architectural, Chemical and Materials, and other related Engineering disciplines.

More information about this series at http://www.springer.com/series/15140

Sanjay Kumar Shukla · Erol Guler
Editors

Advances in Reinforced Soil Structures

Proceedings of the 1st GeoMEast International Congress and Exhibition, Egypt 2017 on Sustainable Civil Infrastructures

Springer

2017 GEOMEAST

SSIGE
SOIL-STRUCTURE INTERACTION GROUP IN EGYPT

Editors
Sanjay Kumar Shukla
School of Engineering
Edith Cowan University
Perth, WA
Australia

Erol Guler
Bogazici University
Istanbul
Turkey

ISSN 2366-3405 ISSN 2366-3413 (electronic)
Sustainable Civil Infrastructures
ISBN 978-3-319-63569-9 ISBN 978-3-319-63570-5 (eBook)
DOI 10.1007/978-3-319-63570-5

Library of Congress Control Number: 2017946436

Printed on acid-free paper

This Springer imprint is published by Springer Nature
The registered company is Springer International Publishing AG
The registered company address is: Gewerbestrasse 11, 6330 Cham, Switzerland

Preface

Toward building sustainable and longer civil infrastructures, the engineering community around the globe continues undertaking research and development to improve existing design, modeling, and analytical capability. Such initiatives are also the core mission of the Soil-Structure Interaction Group in Egypt (SSIGE) to contribute to the ongoing research toward sustainable infrastructure. This conference series "GeoMEast International Congress and Exhibition" is one of these initiatives.

Ancient peoples built their structures to withstand the test of time. If we think in the same way, our current projects will be a heritage for future generations. In this context, an urgent need has quickly motivated the SSIGE and its friends around the globe to start a new congress series that can bring together researchers and practitioners to pursue "Sustainable Civil Infrastructures." The GeoMEast 2017 is a unique forum in the Middle East and Africa that transfers from the innovation in research into the practical wisdom to serve directly the practitioners of the industry.

More than eight hundred abstracts were received for the first edition of this conference series "GeoMEast 2017" in response to the Call for Papers. The abstracts were reviewed by the Organizing and Scientific Committees. All papers were reviewed following the same procedure and at the same technical standards of practice of the TRB, ASCE, ICE, ISSMGE, IGS, IAEG, DFI, ISAP, ISCP, ITA, ISHMII, PDCA, IUGS, ICC, and other professional organizations who have supported the technical program of the GeoMEast 2017. All papers received a minimum of two full reviews coordinated by various track chairs and supervised by the volumes editors through the Editorial Manager of the SUCI "Sustainable Civil Infrastructure" book series. As a result, 15 volumes have been formed of the final +320 accepted papers. The authors of the accepted papers have addressed all the comments of the reviewers to the satisfaction of the track chairs, the volumes editors, and the proceedings editor. It is hoped that readers of this proceedings of the GeoMEast 2017 will be stimulated and inspired by the wide range of papers written by a distinguished group of national and international authors.

Publication of this quality of technical papers would not have been possible without the dedication and professionalism of the anonymous papers reviewers. The names of these reviewers appear in the acknowledgment that follows. For any additional reviewers whose names were inadvertently missed, we offer our sincere apologies.

We are thankful to Dr. Hany Farouk Shehata, Dr. Nabil Khelifi, Dr. Khalid M. ElZahaby, Dr. Mohamed F. Shehata, and to all the distinguished volumes editors of the proceedings of the GeoMEast 2017. Appreciation is extended to the authors and track chairs for their significant contributions. Thanks are also extended to Springer for their coordination and enthusiastic support to this conference. The editors acknowledge the assistance of Ms. Janet Sterritt-Brunner and Mr. Arulmurugan Venkatasalam in the final production of the 15 edited volumes of "Proceedings of GeoMEast 2017".

Contents

Direct Shear Testing of Sand – Geotextile Interfaces

Ioannis N. Markou(✉)

Democritus University of Thrace, Xanthi, Greece
imarkou@civil.duth.gr

Abstract. The effect of shear box size, geotextile type and properties and sand grain shape and size on the sand – geotextile interaction was investigated experimentally by conducting interface tests with conventional and large-scale direct shear equipment. Four clean uniform sands, one with subangular grains and three with rounded grains of different sizes were tested in dry and dense condition. Seven non-woven polypropylene geotextiles of various types and properties and seven woven geotextiles with or without apertures were used in the tests. The large-scale direct shear tests were conducted according to ASTM Standard D5321, using a 300 mm square shear box. The tests with the conventional 100 mm square shear box were performed using the same normal stresses and comparable shearing rate with the large-scale tests. The interface friction coefficient values obtained from the tests with the conventional and the large-scale shear box are in good agreement. Therefore, the 100 mm shear box is satisfactory for testing materials like those used in the present investigation. The values of friction coefficient, *tanδ*, at the sand – geotextile interface are affected by the geotextile type and range from 71% to 104% with respect to the internal friction coefficients, *tanφ*, of sands. Although the interface friction coefficient values are larger in the sand with subangular grains, the efficiency (*tanδ*/*tanφ* ratio) values for the sand with subangular grains are lower in comparison with the sand with rounded grains of equal size. For interfaces between sands of different grain sizes and geotextiles without apertures, the obtained efficiency values generally increase with decreasing sand grain size.

1 Introduction

Design procedures for reinforced sand structures require quantification of the interaction behavior at the sand – reinforcement interface. This is accomplished by conducting large-scale laboratory direct shear and pull-out tests and is expressed in terms of an apparent friction angle, *δ*, or an interface friction coefficient, *tanδ*. The abovementioned experimental procedures are rather costly because they require the use of specially designed and constructed large-size direct shear or pull-out boxes and specialized personnel. More specifically, both ASTM D5321 (2006) and EN ISO 12957-1 (2005) direct shear tests call for a square shear box of 300 mm in size. While such a large shear box is appropriate for geonets, geogrids, many geocomposites, and large particle-sized soils, Koerner (2005) considers it to be excessive for geotextiles (and certainly for geomembranes) against sands, silts and clays. Conventional geotechnical engineering laboratory shear boxes (e.g., 100 mm), are felt to be satisfactory for

S.K. Shukla and E. Guler (eds.), *Advances in Reinforced Soil Structures*,
Sustainable Civil Infrastructures, DOI 10.1007/978-3-319-63570-5_1

geotextile testing (Koerner 2005). It is, therefore, of merit to verify the suitability of interface direct shear tests with 100 mm shear box, by comparing their results obtained for geotextiles manufactured with different processes and having different properties with those of standardized tests with 300 mm shear box.

The interaction behavior at sand – geotextile interfaces has been investigated extensively by conducting direct shear tests and it was found that it depends on the surface characteristics, the type, the strength and the stiffness of the geosynthetic (Williams and Houlihan 1987, Koutsourais et al. 1998). On the other hand, the results of direct shear tests on interfaces between dense Ottawa 20–30 sand and a non-woven needle-punched geotextile of four different densities indicated that the apparent friction angle is independent of the geotextile density (Athanasopoulos et al. 2002). Also, the friction coefficient for rounded sand – woven geotextile interfaces is lower than the one obtained for angular sand – woven geotextile interfaces (Anubhav and Basudhar 2013), but the apparent friction angle can decrease with increasing (Formazin and Batereau 1985) or decreasing (Choudhary and Krishna 2016) sand grain size. The aforementioned observations show that the effect of geotextile type and properties and sand grain shape and size on the sand – geotextile interaction needs further documentation. Toward these ends, 93 direct shear tests with conventional and large-scale equipment were performed on 22 interfaces between sands differing in grain shape or grain size and various woven and non-woven geotextiles, and the results obtained are reported herein.

2 Materials

The direct shear tests were conducted using four clean, uniform sands in dry and dense condition. From the properties of sands presented in Table 1, it can be seen that three of them (designated as R 20–30, R 30–40 and R 40–100) are standard Ottawa quartz sands with rounded grains of different sizes, since their grain sizes are limited between ASTM sieve sizes Nos. 20 and 30, 30 and 40, and 40 and 100, respectively. As shown in Fig. 1, the fourth sand (designated as S 20–30) has subangular grains of the same sizes with R 20–30 sand and was tested in order to investigate the effect of grain shape on sand – geotextile interaction. The values of angle of internal friction, φ, of the sands in dry and dense condition, were determined by conducting triaxial compression tests and are also shown in Table 1. These φ values are used for normalizing the obtained values of the interface friction angle, δ.

Table 1. Sand properties

Sand	Grain shape	Grain sizes (mm)			Void ratios		Shear strength	
		D_{max}	D_{50}	D_{min}	e_{max}	e_{min}	φ (°)	D_r (%)
S 20–30	Subangular	0.85	0.71	0.60	0.96	0.62	47.0	83
R 20–30	Rounded	0.85	0.71	0.60	0.77	0.46	36.0	82
R 30–40	Rounded	0.60	0.51	0.43	0.85	0.52	35.0	92
R 40–100	Rounded	0.43	0.25	0.15	0.79	0.52	37.0	90

S 20-30 (Subangular) Sand **R 20-30 (Rounded) Sand**

Fig. 1. Enlarged views of sands used in study

Seven non-woven polypropylene geotextiles and seven woven geotextiles with or without apertures, provided by eight different manufacturers, were tested during this investigation. The geotextiles were selected in order to cover a wide range of types of the commercially available products and pertinent properties of them, according to the manufacturers, are presented in Table 2. More specifically, two needle-punched with

Table 2. Geotextile properties

Geotextile	Thickness (mm)	Mass per unit area (g/m^2)	Apparent opening size (μm)	Tensile test results	
				Max. tensile load (kN/m)	Extension at max. load (%)
TS 50 (NW)	1.90	200	110	15.0/15.0 *	75/35 *
B 200 (NW)	2.70	201	100	10.6/12.9 *	88/90 *
F 400 (NW)	1.80	275	75	16.5/17.5 *	52/55 *
SF 40 (NW)	0.45	136	120	8.5	60
SF 56 (NW)	0.54	190	80	12.8	65
SF 77 (NW)	0.65	260	60	20.0	70
SF 111 (NW)	0.85	375	55	29.0	70
TP 240 (W)	1.17	240	200	50.0/50.0 *	15/13 *
TP 310 (W)	1.01	310	105	66.0/66.0 *	14/10 *
TP 400 (W)	1.15	400	94	86.0/86.0 *	20/14 *
SG 80/80 (W)	1.35	360	255	82.0/86.0 *	20/11 *
HS 400/50 (W)	1.10	700	—	400.0/50.0 *	<10/<20 *
H 50.145 (W)	1.15	225	—	32.0/32.0 *	15/18 *
N 66447 (W)	0.90	194	1256	44.4/39.6 *	27/22 *

(NW): Non-woven, (W): Woven

*Machine direction/Cross machine direction

comparable properties (POLYFELT TS 50 and BONDEX 200), one needle-punched with thermally treated surfaces (FIBERTEX F 400) and four thermally bonded fabrics with different properties (TYPAR SF 40, SF 56, SF 77 and SF 111), constitute the group of the selected non-woven geotextiles. These materials are designated as TS 50, B 200, F 400, SF 40, SF 56, SF 77 and SF 111, respectively. The set of woven geotextiles consists of three polypropylene with different properties (THRACE PLASTICS 240, 310 and 400), one standard grade polypropylene (BONAR SG 80/80), one high strength polyester/polyamide (HUESKER Stabilenka 400/50), as well as two materials with apertures of different size, one polyester with PVC coating and aperture size, A, equal to 1.20 mm (HUESKER HaTe 50.145) and one polyethylene with aperture size, A, equal to 0.77 mm (NICOLON 66447). The woven geotextiles are designated as TP 240, TP 310, TP 400, SG 80/80, HS 400/50, H 50.145 and N 66447, respectively. Enlarged images of all geotextile types used in the present study are shown in Fig. 2.

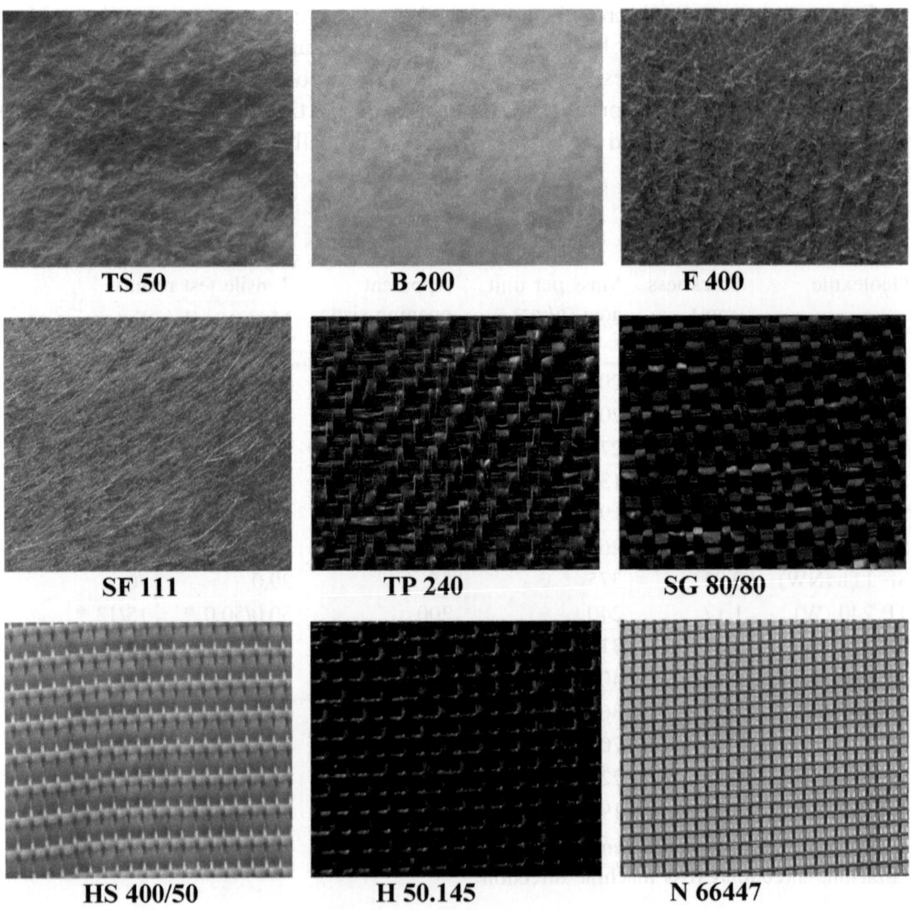

Fig. 2. Enlarged views of geotextiles used in study

3 Experimental Procedures

Conventional and large-scale direct shear equipment was utilized to conduct the tests on sand – geotextile interfaces in order to evaluate the interface friction coefficient, *tanδ*. The direct shear tests with the large shear box were performed on interfaces between dry, dense R 20–30 sand and selected non-woven and woven geotextiles of various types, with the purpose of investigating the effect of shear box size on the sand – geotextile interaction.

The large-scale tests were conducted using a direct shear apparatus of controlled displacement with a 300 mm square shear box. A cross section of the square shear box is shown in Fig. 3a. The normal load, with maximum value of 100 kN, is applied hydraulically to the top plate whereas the horizontal displacement of adjustable rate is applied through electric motors to the lower part of the shear box. The upper part of the shear box is held in place by the reaction of the load ring. A system described in detail by Athanasopoulos et al. (2002), which was designed and fabricated in order to make the shear box capable of accommodating interface shear testing, was used in the tests. For sand – geotextile interface testing, the proper parts of the system were placed and assembled in the lower part of the shear box in the order indicated in Fig. 3b. The geotextile sheet was placed and clamped on the rough interface plate and dry sand was placed and compacted in layers in the upper part of the shear box. The sand was compacted using a hand operated tamper and care was taken in order to produce sand layers with constant density. The relative density of the sand in these tests ranged from 83% to 93%. The reported difference between the angles of internal friction, φ, of loose (D_r = 46%) and dense (D_r = 80%) sand is 6.5° (Gourc et al. 1996). The increase of sand relative density from an average value of 51% (loose condition) to an average value of 93% (dense condition) caused an increase of the friction angle, δ, by 6–7° in sand – non-woven geotextile interfaces (Miyamori et al. 1986, Gourc et al. 1996) and by 5–10° in sand – woven geotextile interfaces (Makiuchi and Miyamori 1988). However, negligible differences in the friction angles φ and δ were obtained by Lee and Manjunath (2000) for an increase of sand relative density from 50% to 80%. From all these data it can be concluded that the range of sand relative density in the present

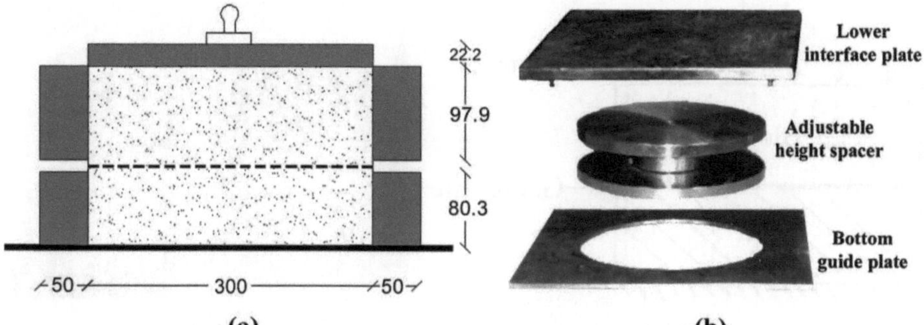

Fig. 3. Large (300 mm) shear box, (a) dimensions, (b) system for sand – geotextile interface testing

study does not have a substantial effect on the results since it is not expected to cause variations in the friction angles φ and δ larger than 1°. The large-scale tests were conducted according to ASTM Standard D 5321 (2006), with normal stresses, σ_n, equal to 100, 200 and 400 kPa, at a constant rate of shearing equal to 1 mm/min and were completed at a horizontal displacement of 30 mm.

All the other tests were conducted using a conventional direct shear apparatus of controlled displacement with a 100 mm square shear box. The specimen configuration used in these tests, is shown schematically in Fig. 4a and is depicted in Fig. 4b. The dry sand was placed and compacted in the lower part of the shear box. The sands were compacted using a hand operated tamper and care was taken in order to produce sand layers with constant density. The geotextile sheet was placed and fixed on the rough surface of a wooden block and, then, the block with the geotextile sheet was placed in the upper part of the shear box in contact with the sand. Taking into consideration that evidence of negative influence on the test results was not found in the literature for this test setup and that it is also suggested by Koerner (2005), it is intuitively believed that placing the sand and the geotextile in the lower and the upper parts of the shear box, respectively, is practically the same as the customary test setup used in the large-scale tests. Since geotextiles of different thickness and compressibility were tested in this study, it was very complicated to adjust the height effectively with the wooden block in the lower part of the shear box so as the sand – geotextile interface to coincide with the shearing plane. This adjustment was accurately made in the large-scale tests with the "adjustable height spacer" included in the system of Fig. 3b. Thus, the specimen configuration of Fig. 4 was preferred in the conventional tests for simplicity reasons. All conventional tests were conducted at a relative density of the sands between 87% and 97%, with normal stresses, σ_n, equal to 100, 200 and 400 kPa, at a constant rate of shearing equal to 0.25 mm/min and were completed after failure at the sand – geotextile interface (peak value of shear force). The chosen rate of shearing is equal to 0.25%/min with reference to the dimension of the shear box in the shearing direction and is comparable to the equivalent shearing rate of 0.33%/min used in the large-scale tests. A number of conventional tests were repeated for the verification of data resulting in differences between shear stress values at failure ranging from 1.1% to 10.6%. Consequently, the repeatability of the tests is considered satisfactory.

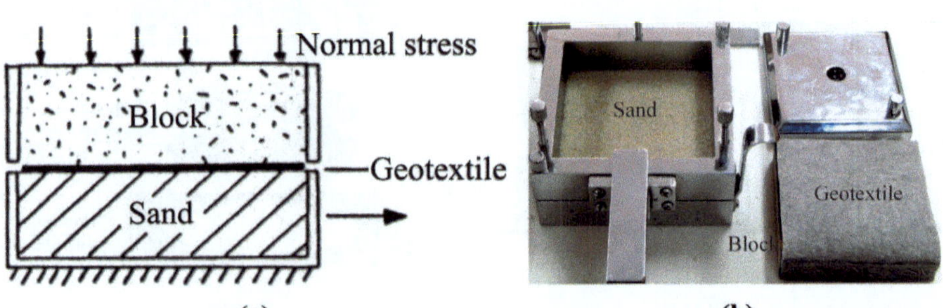

Fig. 4. Specimen configuration for sand – geotextile interface testing with the 100 mm shear box

4 Results and Discussion

Typical "shear stress – horizontal displacement" curves obtained from large-scale and conventional direct shear tests, conducted on various sand – geotextile interfaces, are shown in Fig. 5. All these curves present a peak indicating failure at the sand – geotextile interface. The typical "vertical displacement – horizontal displacement" curves presented in Fig. 6, show an initial decrease and a subsequent increase of the specimen height as shearing progresses. Although the measured values are low, these observations signify compression and expansion at the sand – geotextile interface, respectively. The shear and normal stress values at failure were used to plot the "shear stress – normal stress" diagrams, in order to evaluate the interface shearing resistance between the geotextiles and the sands. As it is typically shown in Fig. 7, the interaction behavior can be described by a linear Mohr – Coulomb failure envelope presenting adhesion values equal to zero. From the slope of the failure envelopes resulted from

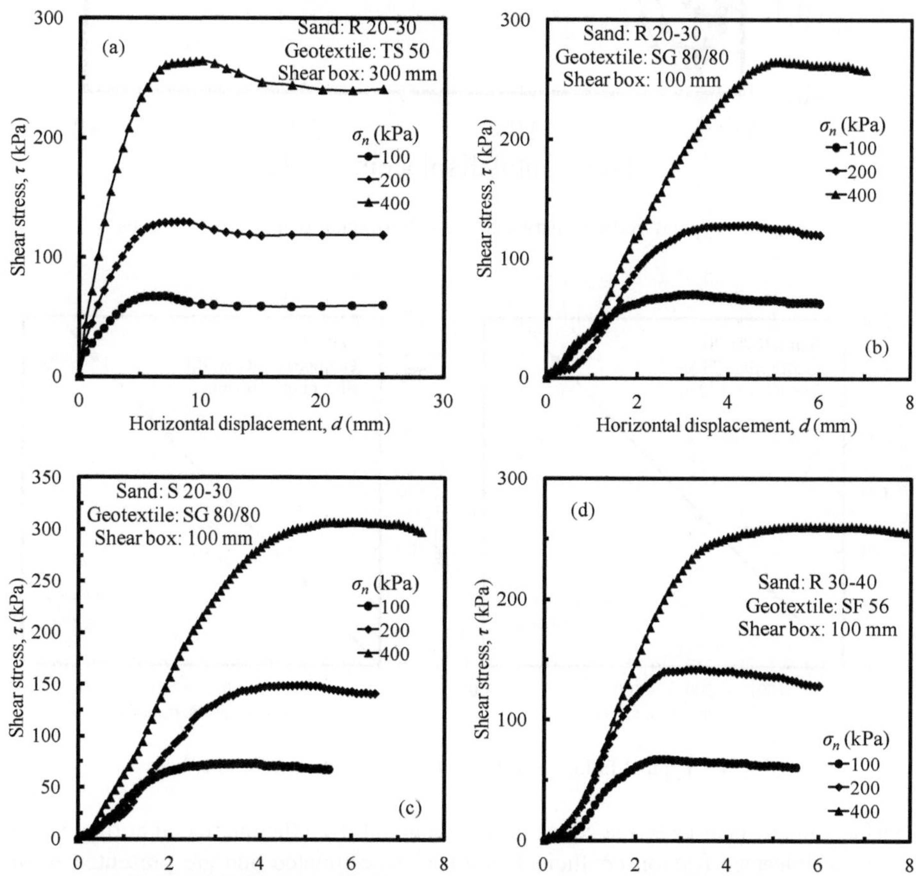

Fig. 5. Typical shear stress – horizontal displacement curves

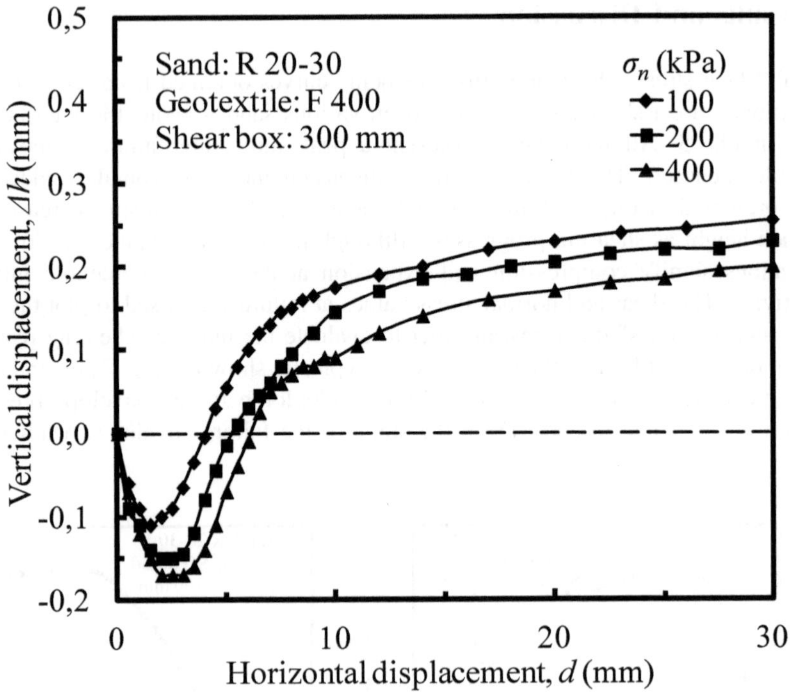

Fig. 6. Typical vertical displacement – horizontal displacement curves

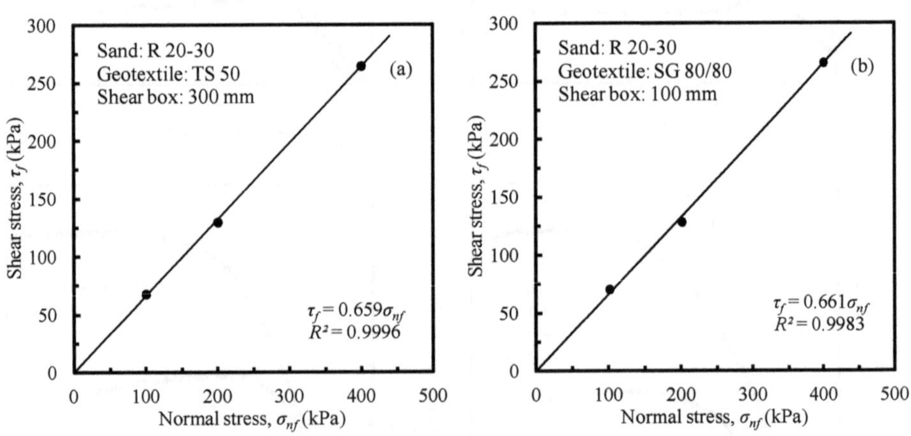

Fig. 7. Typical failure envelopes for sand – geotextile interfaces

interface direct shear tests, the constant (independent from the interfacial normal stress) values of interface friction coefficient, *tanδ*, were estimated and are presented in the following sections. The interface friction coefficient values were also normalized with regard to the internal friction coefficients, *tanφ*, of the corresponding sands.

The resulting values of friction efficiency, $E_{\varphi} = tan\delta/tan\varphi$, range from 71% to 104% and are in good agreement with the typical range of friction property of geotextiles which is equal to 60%–100% of soil friction (Koerner 2005).

Two series of large-scale tests were conducted on the R 20–30 sand – HS 400/50 geotextile interface, one with the shearing direction parallel and one with the shearing direction perpendicular to the production direction of the geotextile. This was dictated by the significant difference in tensile strength of this geotextile in machine and cross machine direction, as shown in Table 2. However, nearly equal *tanδ* values (difference = 2.8%) were obtained for the two directions of HS 400/50 geotextile.

Effect of Shear Box Size

As explained earlier, the direct shear tests with the large-scale and the conventional shear box were conducted using the same normal stresses and equivalent shearing rates with the purpose of comparing their results. This comparison is made in Table 3 for interfaces between dry, dense sand with rounded grains and a variety of woven and non-woven geotextiles. The value of friction coefficient, *tanδ*, from the large-scale tests on HS 400/50 geotextile is the average of the similar values obtained, as stated before, for the two directions of this geotextile. As shown in Table 3, the friction coefficient values resulted from the tests with 300 mm shear box are generally smaller or larger than the ones obtained from the tests with 100 mm shear box. The differences between the *tanδ* values obtained from the two shear boxes can be considered as low, since they are lower than ±6% for the non-woven geotextiles and they range from −10.4% to +13.2% for the woven geotextiles. The larger differences observed for the woven geotextiles can possibly be attributed to the structure and geometry of them. In conclusion, the aforementioned observations indicate that the results of the two tests are comparable and, therefore, that the 100 mm shear box is suitable for interface testing of materials like those used in the present study. For that reason, this conventional shear box was used for the parametric investigation presented in the subsequent sections.

Effect of Geotextile Type and Properties

As also shown in Table 3, the type of geotextile affects substantially the values of interface friction coefficient leading to differences between them as high as 31%.

Table 3. Effect of shear box size and geotextile type on R 20–30 sand – geotextile interface friction

Geotextile	Type	300 mm shear box		100 mm shear box		Difference in *tanδ* *
		tanδ	E_{φ} (%)	*tanδ*	E_{φ} (%)	
B 200	Non-woven	0.69	94	0.69	94	0.0
TS 50	Non-woven	0.66	90	0.63	86	+4.5%
SF 56	Non-woven	0.58	79	0.60	82	−3.4%
F 400	Non-woven	0.69	94	0.73	100	−5.8%
SG 80/80	Woven	0.76	104	0.66	90	+13.2%
HS 400/50	Woven	0.72	99	0.71	97	+1.4%
N 66447	Woven	0.67	92	0.74	101	−10.4%
H 50.145	Woven	0.63	86	0.69	94	−9.5%

*$[(tan\delta_{300\,mm} - tan\delta_{100\,mm})/tan\delta_{300\,mm}] \cdot 100$

Table 4. Effect of geotextile properties on R 20–30 sand – geotextile interface friction

Geotextile	Type	Mass per unit area (g/m^2)	Max. tensile load (kN/m)	Coefficient of friction *tanδ*	Efficiency E_φ (%)
SF 40	Non-woven	136	8.5	0.62	85
SF 56	Non-woven	190	12.8	0.60	82
SF 77	Non-woven	260	20.0	0.63	86
SF 111	Non-woven	375	29.0	0.57	78
TP 240	Woven	240	50.0/50.0	0.60	82
TP 310	Woven	310	66.0/66.0	0.57	78
TP 400	Woven	400	86.0/86.0	0.60	82

However, non-woven geotextiles generally present *tanδ* values in the same range as woven geotextiles. Considering the properties of non-woven or woven geotextiles of the same types and manufacturers (Table 4), the *tanδ* value does not present a consistent variation with increasing mass per unit area and tensile strength of the geotextile. The abovementioned observations indicate that the sand – geotextile interaction behavior depends mainly on the surface characteristics of the geotextiles which are strongly influenced by the geotextile type.

Effect of Sand Grain Shape and Size

The results of direct shear tests conducted with four different geotextiles and two sands having the same grain size and differing in grain shape are shown in Table 5. It is easily observed that the values of friction coefficient are higher in sand with subangular grains than in sand with rounded grains. On the contrary, the friction efficiencies, E_φ, are higher in the sand with rounded grains indicating a more effective mobilization of soil friction in comparison with the sand with subangular grains. The same trend is also noticed on the basis of the results reported by Anubhav and Basudhar (2013) for two woven geotextiles in contact with one rounded and one angular particle sand, having φ values equal to those of the sands tested in the present study. This behavior is attributed to the larger difference between the *tanφ* values compared to the difference between the *tanδ* values of the sands differing in grain shape. Presented in Table 6 are the results of direct shear tests conducted with one non-woven and one woven geotextile without apertures in contact with three sands having the same (rounded) grain shape and differing in grain size. Although an increase of friction coefficient with decreasing sand grain size is obvious only for the non-woven geotextile, the friction efficiency,

Table 5. Effect of sand grain shape on sand – geotextile interface friction

Geotextile	Type	Rounded sand (R 20–30)		Subangular sand (S 20–30)	
		tanδ	E_φ (%)	*tanδ*	E_φ (%)
SF 56	Non-woven	0.60	82	0.78	73
SG 80/80	Woven	0.66	90	0.76	71
H 50.145	Woven	0.69	94	0.92	86
N 66447	Woven	0.74	101	0.82	77

Table 6. Effect of sand grain size on sand – geotextile interface friction

Rounded sand	Grain size D_{50} (mm)	Non-woven geotextile (SF 56)		Woven geotextile (SG 80/80)	
		$tan\delta$	E_{φ} (%)	$tan\delta$	E_{φ} (%)
R 20–30	0.71	0.60	82	0.66	90
R 30–40	0.51	0.66	94	0.67	96
R 40–100	0.25	0.72	96	0.66	88

$tan\delta/tan\varphi$, increases with decreasing sand grain size in both geotextiles with the exception of R 40–100 sand – SG 80/80 geotextile interface. This increase may be possibly attributed to the more efficient mobilization of soil friction by the larger number of grains in contact with the geotextile, as the sand grain size decreases.

5 Conclusions

Based on the results of this investigation and within the limitations posed by the number of tests conducted and the materials used, the following conclusions may be advanced:

- The interaction at the sand – geotextile interface can be described by linear failure envelopes presenting negligible adhesion values and friction coefficient values ranging from 71% to 104% with regard to the internal friction coefficients of sands.
- The direct shear tests conducted with 300 mm and 100 mm square shear boxes gave comparable values of interface friction coefficient. Therefore, the 100 mm shear box is appropriate for interface testing of materials similar to those used in the present investigation.
- The sand – geotextile interaction behavior depends on the surface characteristics of the geotextiles which are strongly influenced by the geotextile type.
- The rounded shape and the size decrease of sand grains were found to mobilize more effectively the soil friction at the sand – geotextile interface.
- The aforesaid conclusions are limited to uniform sands with grain sizes similar to those used in this investigation. Also, the effect of well graded backfill materials on the soil – geotextile interaction is not discussed in the present study.

Acknowledgments. The interface direct shear tests with the 300 mm shear box were conducted by the author in the University of Patras, Greece (Department of Civil Engineering, Geotechnical Engineering Laboratory). Thanks are expressed to Professor D.K. Atmatzidis for the permission to use this equipment for conducting the tests. The interface direct shear tests with the 100 mm shear box were conducted in the Soil Mechanics & Foundation Engineering Laboratory of Democritus University of Thrace by the students P. Aggonas and D. Ioannou, whose careful work is gratefully acknowledged.

References

Anubhav, Basudhar, P.K.: Interface behavior of woven geotextile with rounded and angular particle sand. J. Mater. Civil Eng. **25**(12), 1970–1974 (2013). doi:10.1061/(ASCE)MT.1943-5533.0000774

ASTM D5321: Standard test method for determining the coefficient of soil and geosynthetic or geosynthetic and geosynthetic friction by the direct shear method. Geosynthetics, vol. 04.13. ASTM International, West Conshohocken (2006)

Athanasopoulos, G.A., et al.: Evaluation of sand – geotextile interface friction angle by a modified 300 x 300 mm direct shear box. In: Seventh International Conference on Geosynthetics, Nice, France, vol. 4, pp. 1301–1304 (2002)

Choudhary, A.K., Krishna, A.M.: Experimental investigation of interface behaviour of different types of granular soil/geosynthetics. Int. J. Geosynth. Ground Eng. **2**(4), 11 p. (2016). doi:10.1007/s40891-016-0044-8

EN ISO 12957-1: Geosynthetics – Determination of friction characteristics – Part 1: Direct shear test. European Committee for Standardization, Brussels, Belgium (2005)

Formazin, J., Batereau, C.: The shear strength behavior of certain materials on the surface of geotextiles. In: XIth International Conference on Soil Mechanics and Foundation Engineering, San Francisco, USA, vol. 3, pp. 1773–1775 (1985)

Gourc, J.P., et al.: Friction measurement by direct shearing or tilting process – development of a European standard. In: First European Geosynthetics Conference, Maastricht, Netherlands, pp. 1039–1046 (1996)

Koerner, R.M.: Designing with Geosynthetics, 5th edn. Pearson – Prentice Hall, Upper Saddle River (2005)

Koutsourais, M., et al.: Soil interaction characteristics of geotextiles and geogrids. In: Sixth International Conference on Geosynthetics, Atlanta, USA, vol. 2, pp. 739–744 (1998)

Lee, K.M., Manjunath, V.R.: Soil – geotextile interface friction by direct shear tests. Can. Geotech. J. **37**(1), 238–252 (2000)

Makiuchi, K., Miyamori, T.: Mobilization of soil – geofabric interface friction. In: International Geotechnical Symposium on Theory and Practice of Earth Reinforcement, Fukuoka, Japan, pp. 129–134 (1988)

Miyamori, T., et al.: Frictional characteristics of non-woven fabrics. In: Third International Conference on Geotextiles, Vienna, Austria, vol. 3, pp. 701–705 (1986)

Williams, N.D., Houlihan, M.F.: Evaluation of interface friction properties between geosynthetics and soils. In: Geosynthetics 1987, New Orleans, USA, vol. 2, pp. 616–627 (1987)

Consideration of Geosynthetic Tension in Interpretation of Data from Inclined Plane Tests

Atef Ben Othmen(✉) and Mounir Bouassida

Université de Tunis El Manar, École Nationale d'Ingénieurs de Tunis, LR14ES03, Ingénierie Géotechnique, BP 37 Le Belvédère, 1002 Tunis, Tunisia
atef.othman@gmail.com

Abstract. The inclined plane device is specifically adapted to assess the geosynthetic interfaces friction under low confinement conditions. The lack of the standardized procedure proposed by the European standard (EN ISO 12957-2 2005) for the determination of geosynthetic interface friction properties has been proven by many recent researches available in the literature.

These researches demonstrated the need for revising the standard displacement procedure since it seems to be poorly suited for many geosynthetics interfaces and because the high sensitivity of the determined friction angle to test conditions.

Geosynthetics of reinforcement interface properties were determined by carrying out inclined plane tests under low confinement adapted to landfill covers conditions. Interface friction angles $\left(\varphi^{stand}\right)$ were determined conformingly to the standardized displacement procedure and compared with those defined by a method known from the literature.

Then, interface friction angles were determined according to a new method called "tension procedure" which considers the measurement of the tension developed by the geosynthetic reinforcement during inclined plane tests.

Compared to previous data, the proposed method allows a new and more comprehensive interpretation of the inclined plane test since it pays attention to the different behavior of geosynthetics during tests.

This research demonstrates that the consideration of reinforcement tensions allows a more suitable comprehension of the mechanical behavior of soil-geosynthetic interface and a better representation of the in-situ behavior of the geosynthetic interface.

It appears that the mechanical behavior of the geosynthetic of reinforcement and the magnitude of the measured tension depended on the reinforcement characteristics specially the geosynthetic surface structure and the tensile stiffness (tensile modulus) *J* (*kN/m*).

The determined friction angles, both in static $\left(\varphi^{stat}\right)$ and dynamic $\left(\varphi^{dyn}\right)$ inclined plane conditions, are lower than those calculated using the previous methods which allow a more accurate design of landfill cover systems.

S.K. Shukla and E. Guler (eds.), *Advances in Reinforced Soil Structures*, Sustainable Civil Infrastructures, DOI 10.1007/978-3-319-63570-5_2

1 Introduction

Geosynthetic interface friction angles can be determined using the modified direct shear box test and the inclined plane test following the European standard *EN ISO 12957-2 (*2005*)*.

Many recent studies indicated that the inclined plane is the more appropriate device for the measurement of geosynthetic interface friction angles under low normal stress (Gourc and Reyes-Ramirez 2004; Briançon et al. 2011; Stoltz et al. 2012; Carbone 2013).

However, these studies showed that friction angles measured following the European standard method is non conservative and suggested a revision of the standard testing procedure.

In fact, Gourc and Reyes-Ramirez (2004) demonstrated that standard friction angle φ^{stand} was assessed from a static analysis for conditions that were actually dynamic and proposed a "static – dynamic" procedure which defined a static φ^{stat} and a dynamic φ^{Dyn} friction angles.

The dynamic friction angle φ^{Dyn} was calculated taking into account the displacement acceleration γ of the upper geosynthetic during the non- stabilized displacement on the inclined plane.

However, it was difficult to determine graphically the acceleration γ especially for the interfaces involving geosynthetics having surfaces with significantly sized apertures and questions raised about the effect of the dynamic conditions on the interface properties in such cases.

Briançon et al. (2011) proposed a new method called "force procedure" where a cable connected the soil retaining box to a force sensor fixed on the inclined plane.

Stoltz et al. (2012) proposed a light modification of the "force procedure" by substituting the cable with a spring connecting the soil retaining box and the device frame.

The proposed procedure, called "residual friction procedure", allows the determination of the residual friction properties of geosynthetic interfaces in almost static displacement conditions.

The main disadvantage of both the "force procedure" and the "residual friction procedure" is that it allows only the determination of the residual interface properties φ^{res} which have an importance only in situations where failure may result due to large movement along the soil - geosynthetic interface. This situation is unlikely common in site conditions.

The prime importance in geosynthetic interface design should be given to static and dynamic interface friction angles since small displacements generally occur along the potential failure geosynthetic interface.

In order to assess the different geosynthetic interface friction angles at various kinematic conditions and to understand the temperature influence on the interface friction, Carbone (2013) conducted inclined plane tests according to a new test procedure called "unified inclined plane procedure".

This method is based on grouping the "static – dynamic procedure" of Gourc and Reyes-Ramirez (2004) and the "force procedure" of Briançon et al. (2011) in one unified procedure.

The "unified inclined plane procedure" seems to be a more suitable method since it allows the assessment of different interface friction angles in different kinematic conditions (static, dynamic and residual).

However, as well as the "static – dynamic procedure", this method remains unable to explain the difference between the transition phase extensions as recorded during the carried out tests;and it does not consider the mechanical characteristics of the tested geosynthetic when the geosynthetic interface friction angles are calculated.

From all these studies, it appears that the suggestion of a more accurate testing procedure for the inclined plane tests which allows a more precise measurement of geosynthetic interface friction angles is actually a pending question.

In this paper, all of the tests were conducted using an inclined plane available in the LTHE laboratory of Grenoble (France) which is designed in accordance with the European project of standardization (*EC Measurement and Testing Program Project 0169 –Task 3.2: Friction – 1996, 1997*).

Herein, four soil-geosynthetic of reinforcement interfaces were tested using the inclined plane in order to choose the most performing product in soil stabilization among two geotextiles of reinforcement and two geocomposites obtained by gluing two geomats to the previous geotextiles.

Interface friction angles were firstly calculated according the European standard method (*EN ISO 12957-2* 2005), and then by using the "static- dynamic" method of Gourc and Reyes-Ramirez (2004). The main disadvantages of these methods were discussed.

Furthermore, the interface friction angles were recalculated using a proposed method, called "tension procedure", which considers the reinforcement tension as a resistant force to the tangential shear stress along the slope.

2 Test Apparatus and Materials

Experimental tests were carried out using an inclined plane designed for characterizing the interaction mechanism at soil – geosynthetic interfaces (Fig. 1).

For each geosynthetic reinforcement tested, a sample of geosynthetic (*0.8* m *1.3* m) is fixed to the top of the inclined plane and filled with sand.

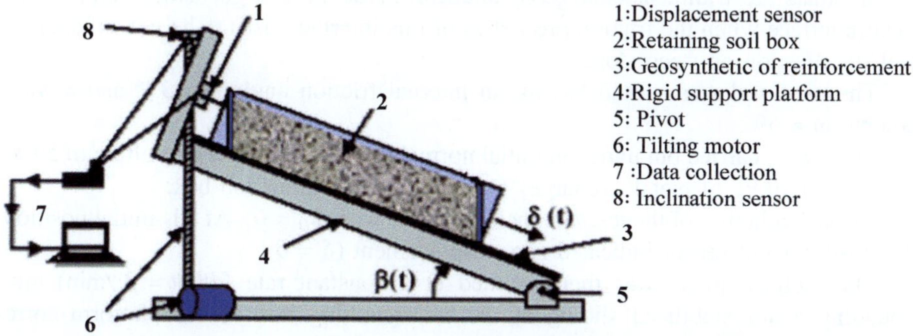

Fig. 1. Inclined plane device

A soil retaining box (Plexiglas) is placed on the geosynthetic sample and filled with 5 cm thick layer of soil and steel plates (metallic charges) to provide the initial normal stress on the soil-geosynthetic interface.

During the tests, the box slides along the plane on two guiding poles fixed at the both sides of the inclined plane. The contact between the guiding poles and the soil retaining box is assumed to be frictionless.

A displacement sensor, fixed to the rigid support of the plane and relied to the soil retaining box by a cable, measured the displacement of the box as the inclination angle of the plane grows.

Tests results were recorded by computer system and saved in an "xls" file format which registered data of inclination angle, soil retaining box displacement and time of test.

Four geosynthetic reinforcement materials have been tested (Fig. 2):

- Non – woven geotextile, reinforced with polyester fibers (Gtr_1).
- Woven geotextile, reinforced with black polyester fibers (Gtr_2).
- Geocomposite [Gtr_1-$Gmat_1$]: a polypropylenegeomat with *8* mm length fibers is glued on the geotextile Gtr_1.
- Geocomposite [Gtr_2-$Gmat_2$]: a polyester geomat with *6* cm length fibers is textured on the geotextile Gtr_2.

Geotextile Gtr_1 Geocomposite[Gtr_1-$Gmat_1$] Geotextile Gtr_2 Geocomposite[Gtr_2-$Gmat_2$]

Fig. 2. Geosynthetic materials tested

Geomats are tridimensional geosyntheticsused at the soil-geotextile of reinforcement interface when the friction properties of this interface risks to be not sufficient in soil stabilization on steep slopes.

The used soil was a sand having an internal friction angle $\varphi = 35°$ and a water content $\omega = 6\%$.

Tests were carried out under an initial normal stress $\sigma'_0 = 7.5$ kPa resulting of *53* kg of soil and *40* kg of metallic charges placed in the retaining soil box.

At the beginning of the test, the inclined plane was ($\beta_0 = 0$). At this initial position, the displacement sensor indicated zero displacement ($\delta = 0$).

The inclined plane was then inclined at a constant rate ($d\beta/dt = 3°$/min) until obtaining a non-stabilized sliding of the soil retaining box on the platform corresponding to an inclination angle (β_s).

3 Test Results

Result repeatability was verified for each test and only the representative values of the inclination angles was used to calculate the interface friction angles.

For each soil – geosynthetic interface tested, results were presented as displacement (δ) versus slope angle (β) curve.

The ISO standard *12957-2* evaluates the interface friction angles $\left(\varphi^{stand}\right)$ for a sliding displacement $\delta = 50$ mm and for the following mechanical diagram (Fig. 3).

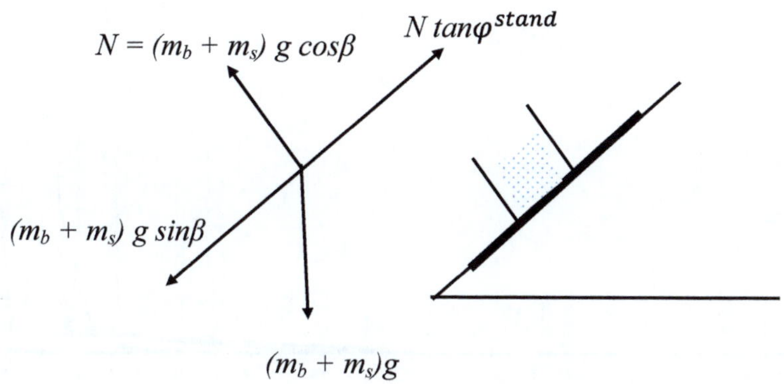

Fig. 3. Forces acting on the soil-geosynthetic interface during the inclined plane test.

The interface friction angles were calculated for an inclination angle β_{50} corresponding to a box displacement of *50* mm using the equation:

$$tan\ \varphi^{stand} = \frac{(m_b + m_s) \cdot g \cdot sin\ \beta_{50}}{m_s.g.cos\ \beta_{50}} \tag{1}$$

Table 1 summarizes the φ^{stand} values of the four tested soil – geosynthetic of reinforcement interfaces. Calculations were done using the following parameters:

m_b: soil retaining box mass (*28.8* kg).
m_s: metallic charges and soil masses (*93* kg).
N: normal reaction of the interface.

The placement of a geomat in the soil-geotextile interface improved considerably the interface friction since friction angles increases six (*6*) degrees for the first geotextile (Gtr_1) and height (*8*) degrees for the second geotextile (Gtr_2).

Table 1. φ^{stand} values determined for the tested interfaces

Structure de renforcement	β_{50} (°)	ϕ^{stand} (°)
Gtr_1	18	26.5
$[Gtr_1\text{-}Gmat_1]$	22	32
Gtr_2	17	25
$[Gtr_2\text{-}Gmat_2]$	22.5	33

However, as shown on Figs. 4 and 5, the displacement of the soil retaining box started at an inclination $\beta = 13°$ for the geocomposite [Gtr_1-$Gmat_1$] and $16°$ for the geocomposite [Gtr_2-$Gmat_2$]. This difference is due to the different behavior (deformability) of the tested geocomposites which is not considered by the standard method.

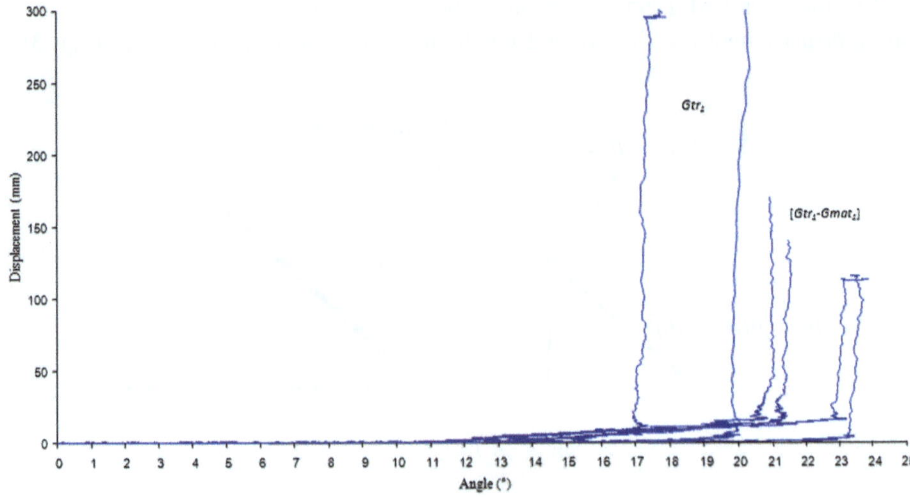

Fig. 4. Horizontal displacement (δ) versus inclination angle β for the interfaces soil - [Gtr_1-$Gmat_1$] and soil - Gtr_1

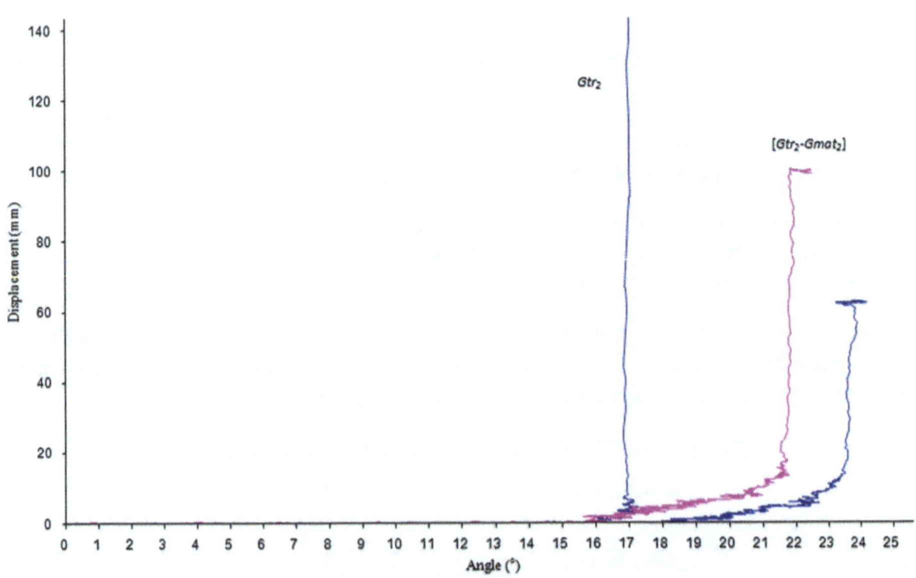

Fig. 5. Horizontal displacement (δ) versus inclination angle β for the interfaces soil - [Gtr_2-$Gmat_2$] and soil - Gtr_2

Moreover, evaluating the interface friction angle ϕ^{stand} for a plane inclination corresponding to a relative sliding displacement $\delta = 50$ mm not give relevant value of the friction angle since test conditions are not static, Gourc et al. (2008).

Consequently, many studies available in the literature suggested a revision of the standard procedure and tried to give a more comprehensive interpretation of the inclined plane test in order to deduce additional information which may be utilized valuably in interface behavior interpretation, (Gourc and Reyes-Ramirez 2004; Briançon et al. 2011; Stoltz et al. 2012 and Carbone 2013).

In the next section, inclined plane tests carried out are reinterpreted in accordance with the "static – dynamic procedure" proposed by Gourc and Reyes-Ramirez (2004) to prove the inadequacy of the current European test standard.

4 "Static-Dynamic Procedure" (Gourc and Reyes-Ramirez 2004)

This interpretation method distinguishes three phases during the inclined plane test (Fig. 6):

- A static phase: during which the soil retaining box is practically immobile ($\delta = 0$). This phase extends from the beginning ($\beta = 0$) of the test until the beginning of the movement of the retaining box over the inclined plane at the inclination angle β_0.
- A transitory phase: during which the soil retaining box move gradually downwards ($\beta_0 < \beta < \beta_s$).

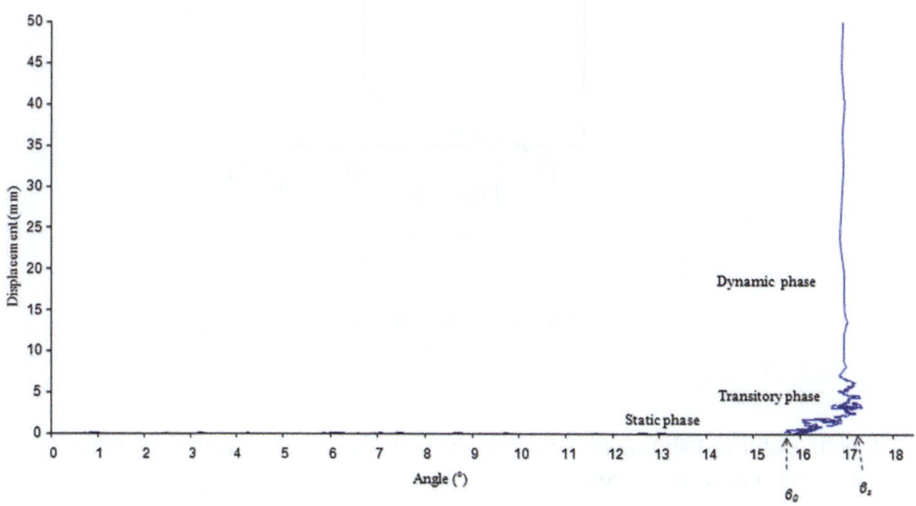

Fig. 6. Phases of the inclined plane test, Gourc et Reyes–Ramirez (2004).

Three types of transitory phases are possible and are function of the sliding mechanism of soil retaining box on the inclined plane:

- A non-stabilized sliding of the soil retaining box with a nonexistent transitory phase ($\beta_0 = \beta_s$): a sudden sliding- abrupt displacement (Fig. 7).

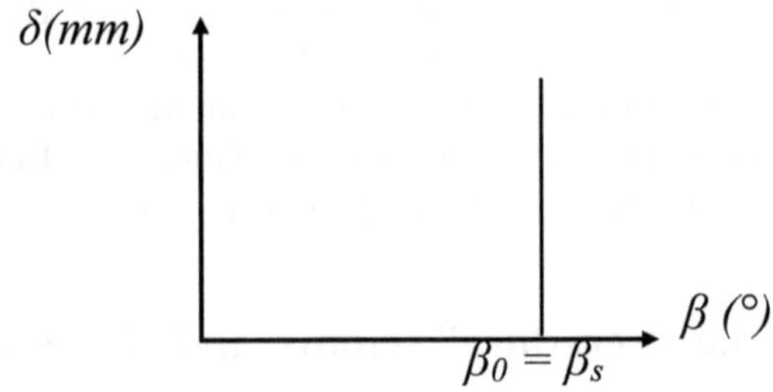

Fig. 7. Non existent transient phase

- A gradual sliding - displacement which increases progressively with the plane inclination (β) (Fig. 8).
- A jerky sliding- displacement increasing in a “stick-slip” fashion (Fig. 9).

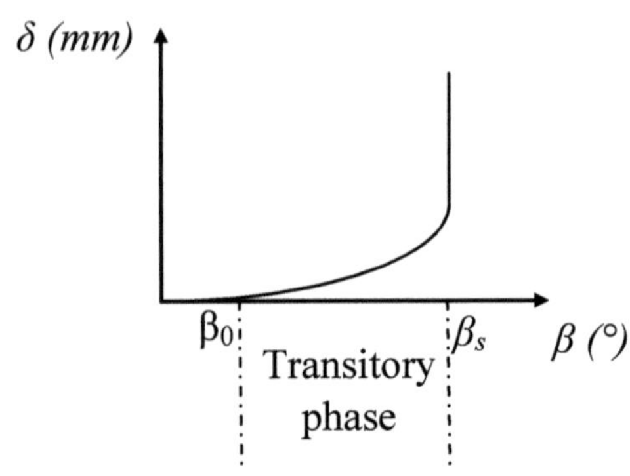

Fig. 8. Gradual sliding.

– A non –stabilized sliding phase where the soil retaining box is in a state movement with a constant acceleration γ.

The authors reported that the standard friction angle was not conservative since the adopted mechanical analysis was not conducted in dynamics although the soil retaining box is in a state of movement at the inclination β_{50} considered by the standard procedure.

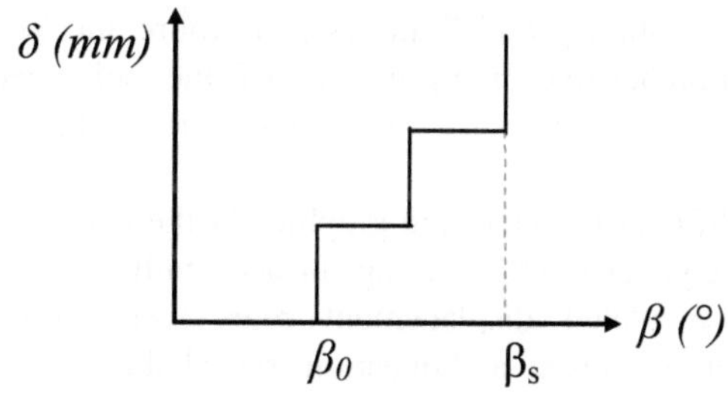

Fig. 9. Jerky sliding.

Consequently, two different interface friction angles were defined:

- Static friction angle ϕ^{stat} corresponding to the initialization of the movement ($\beta_O \leq \beta \leq \beta_s$).

$$tan\,\varphi^{stat} = \frac{(m_b + m_s).g \cdot sin \cdot \beta_0}{m_s \cdot g \cdot cos\,\beta_0} \tag{2}$$

- Dynamic friction angle φ^{dyn} calculated in dynamic conditions when the soil retaining box enters in movement with constant acceleration ($\beta \geq \beta_s$).

The constant acceleration γ was determined graphically using the displacement δ/time t and displacement rate/time t graphs.

$$tan\,\varphi^{dyn} = \frac{(m_s + m_b) \cdot g \cdot sin\,\beta_s - (m_s + m_b)\gamma}{m_s.g.cos\,\beta_s} \tag{3}$$

Static and dynamic friction angles were calculated for the tested soil - geosynthetic interfaces (Table 2).

Table 2. Static and dynamic friction angles of the tested soil - geosynthetic interfaces

Geosynthetic of reinforcement	β_0 (°)	φ^{stat}(°)	β_s (°)	Displacement rate min and max (mm/s)	γ (ms^{-2})	φ^{dyn} (°)
Gtr_1	14	20	17	0.2/2.1	3.45 E-3	23.6
[Gtr_1-$Gmat_1$]	13	19	22	0.1/0.55	1.12 E-4	28.5
Gtr_2	16	23	17	0.2/6.6	1.6 E-3	23.6
[Gtr_2-$Gmat_2$]	16	23	22	0.2/0.62	1 E-4	29.6

The calculated friction angles φ^{stat} and φ^{dyn} were lower than the standard friction angle φ^{stand}.

It appears that the "static-dynamic" analysis of Gourc and Reyes–Ramirez (2004) allowed a more comprehensive interpretation of the soil -geosynthetic interface behavior and the determined interface properties were more conservative than the standard friction angle.

However, it was difficult to determine graphically the acceleration γ especially for the interfaces involving geosynthetics having surfaces with significantly sized apertures like geomats. The calculated displacement rates were very low for the two soil-geocomposite interfaces tested and questions raised about the effect of the dynamic conditions on the interface properties in such cases.

Moreover, different transitory phase extensions were remarked for the four geosynthetic interfaces tested. It extends *6°* for the soil - [*Gtr₂-Gmat₂*] interface and *9°* for the soil - [*Gtr₁-Gmat₁*] interface. This difference was not considered by the "static – dynamic procedure".

5 New Proposed Procedure: "Tension Procedure"

The inclined plane tests described above were carried out under an initial normal stress $\sigma'_0 = 7.5$ kPa.

The same geosynthetic samples were tested under lower initial stresses (*4* and *5* kPa) in order to understand its influence on the tested geosynthetic interface behavior (Fig. 10).

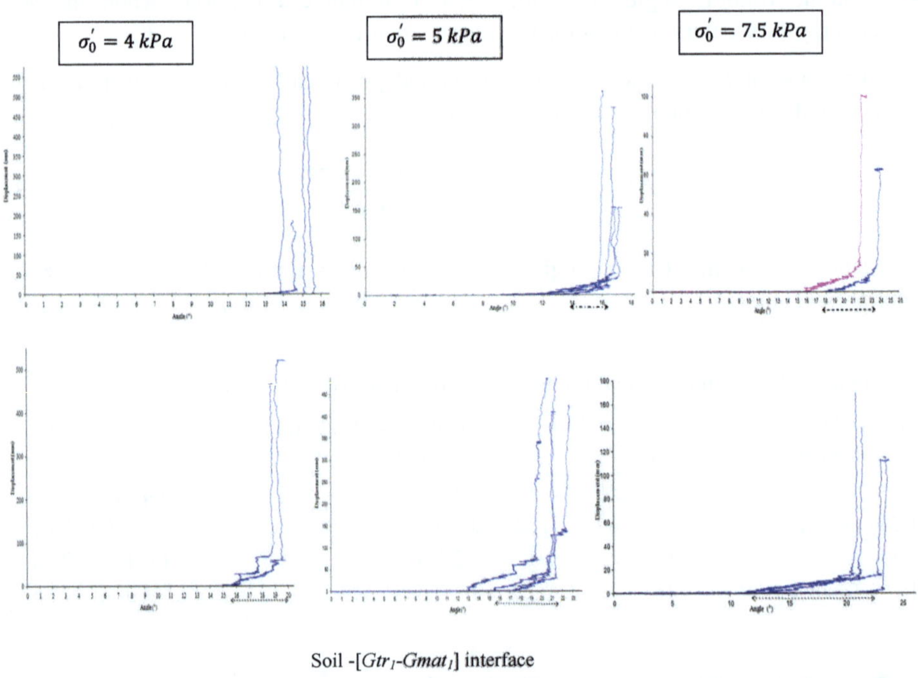

Soil -[*Gtr₁-Gmat₁*] interface

Fig. 10. Influence of the initial normal stress on the transitory phase extension of the tested soil - geocomposite interfaces.

Results displayed in Fig. 10 and Table 3 for the Soil - [Gtr_1-$Gmat_1$] and Soil - [Gtr_2-$Gmat_2$] interfaces indicate that the transitory phase extension increases when increasing the initial normal stress.

Table 3. Variation of the transitory phase extension with the initial normal stress

Initial normal stress (*kPa*)	Geosynthetic reinforcement	
	[Gtr_1-$Gmat_1$]	[Gtr_2-$Gmat_2$]
4	4°	Brutal sliding
5	7°	2°
7.5	12°	5°

It appears that the sliding mechanism (jerky, gradual or sudden) is not a constant behavior of geosynthetic interfaces as reported by Gourc and Reyes-Ramirez (2004) but also depends on the initially applied load on the tested interface and the geometric structure of the geosynthetic surface.

At the beginning of the test where test conditions are static ($0 \leq \beta \leq \beta_0$), the shear strength mobilization depends on the interface roughness.

Higher initial normal stress leads to higher mobilization of shear strength at the soil-geosynthetic interface and within the geosynthetics themselves as a tension transmitted to the geosynthetic anchorage at the top of the inclined plane.

The increase of the plane inclination β leads to an increase of the shear strength ($sin\beta$ increases) and a decrease of the shear resistance ($cos\beta$ decreases). Consequently, the tension mobilized by the geocomposites increases during the static phase of the test.

The mobilized tension F provides a geosynthetic deformation if this geosynthetic of reinforcement have not a sufficient tensile modulus (J)

$$F = J \times \varepsilon \tag{4}$$

Where:

J: tensile modulus of the geosynthetic of reinforcement (*kN/m*).
ε: geosynthetic deformation (dimensionless).

The sudden sliding- abrupt displacement is obtained in two cases:

- The first case when the initial normal stress is low and the low mobilized tension cannot provide the geosynthetic deformation. For $\beta = \beta_s$ the shear strength exceeds the interface shear resistance and the soil retaining box is forced to move on the inclined plane.
- The second case when the geosynthetic tensile modulus is very high and the relatively high mobilized tension cannot provide the geosynthetic deformation although the test is carried out under high initial normal stress.

A transitory phase is obtained from a plane inclination β_0 if the geosynthetic tensile modulus (J) is not sufficient to avoid geosynthetic deformation. In this phase, there is no displacement between the soil and the geosynthetic and test conditions can be considered as pseudo-static.

Consequently, the geosynthetic mobilized tension during an inclined plane test is an important force which should be considered as a resistant effort to the interface shear stress.

The non-consideration of the geosynthetic tension leads to an overestimation of the interface friction angle.

When carrying out the inclined plane tests described above which were conducted under an initial normal stress $\sigma' = 7.5$ kPa, the geosynthetic tension was measured by a force captor installed at the top of the inclined plane (Fig. 11).

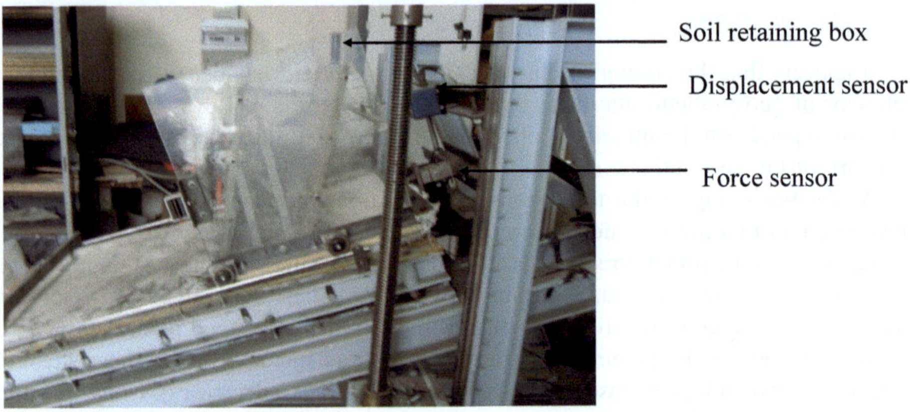

Fig. 11. Measurement of the geosynthetic tension during the inclined plane test.

To assess the geosynthetic mobilized tension during the inclined plane tests, the diagrams geosynthetic tension (F) versus displacement (δ) were presented for the four tested geosynthetic of reinforcement in addition to the classic diagrams displacement (δ) versus inclination (β) (Fig. 12).

Geosynthetic tension (F^{stat}) mobilized during the static phase ($0 < \beta \leq \beta_0$) are presented in Table 4.

Since all tests were conducted under the same initial normal stress, variation of the geosynthetic tension magnitude was a result of different geosynthetic surface roughness which caused a variation of the mobilized static friction.

Consequently, the two geotextiles Gtr_1 and Gtr_2 presented nearly the same surface roughness since it mobilized the same tension while the geocomposite[Gtr_1-$Gmat_1$]was rougher than [Gtr_2-$Gmat_2$].

During the transitory phase, the deformation of the geotextile Gtr_2 is lower than that of the geotextile Gtr_1 although it mobilized more pseudo-dynamic tension F^{psdyn} (Fig. 12).

These different behaviors were occurred due todifferent mechanical characteristics of the two tested geotextiles since Gtr_2 has a higher tensile modulus (J) than that of Gtr_1(*100* kN/m versus *58* kN/m).

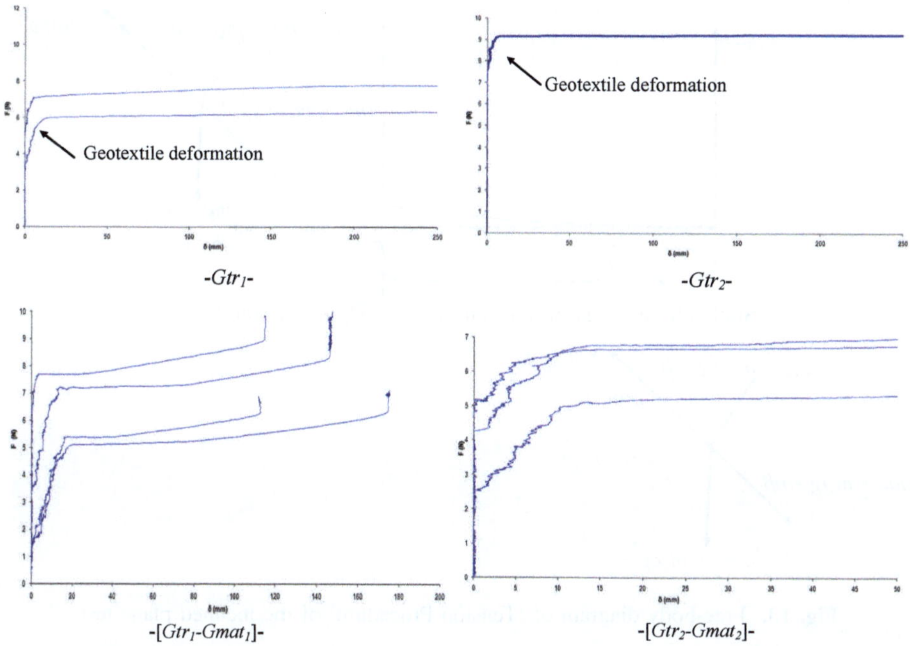

Fig. 12. Geosynthetic tension (*F*) versus displacement of the tested geosynthetics.

Table 4. Geosynthetic tension during the static phase.

Geosynthetic of reinforcement	F^{stat} (N)
Gtr_1	8
Gtr_2	9
$[Gtr_1\text{-}Gmat_1]$	6
$[Gtr_2\text{-}Gmat_2]$	4.5

The deformation of the geocomposite $[Gtr_1\text{-}Gmat_1]$ started earlier than the deformation of $[Gtr_2\text{-}Gmat_2]$ ($\beta_0 = 12°$ instead $\beta_0 = 16°$). It was a result of higher tensile modulus (*J*) of the geocomposite$[Gtr_2\text{-}Gmat_2]$ which allowed a longer static phase.

During the dynamic phase ($\beta_0 < \beta \leq \beta_s$), the geocomposite $[Gtr_1\text{-}Gmat_1]$ continued the deformation caused by a dynamic friction mobilization within the tangled fibers of the geomat. These fibers were loose during the static phase where the tension was mobilized only by the geotextile Gtr_1.

However, the behavior of the geocomposite $[Gtr_2\text{-}Gmat_2]$ was different since its geomat was made of relatively shorter fibers textured on the geotextile Gtr_2. This geocomposite exhibited the same tension mobilized during the transitory phase without additional deformation.

Interface friction angles were recalculated by considering the mobilized tensions during the different phases of the inclined plane tests in accordance with the following free-body diagram (Fig. 13).

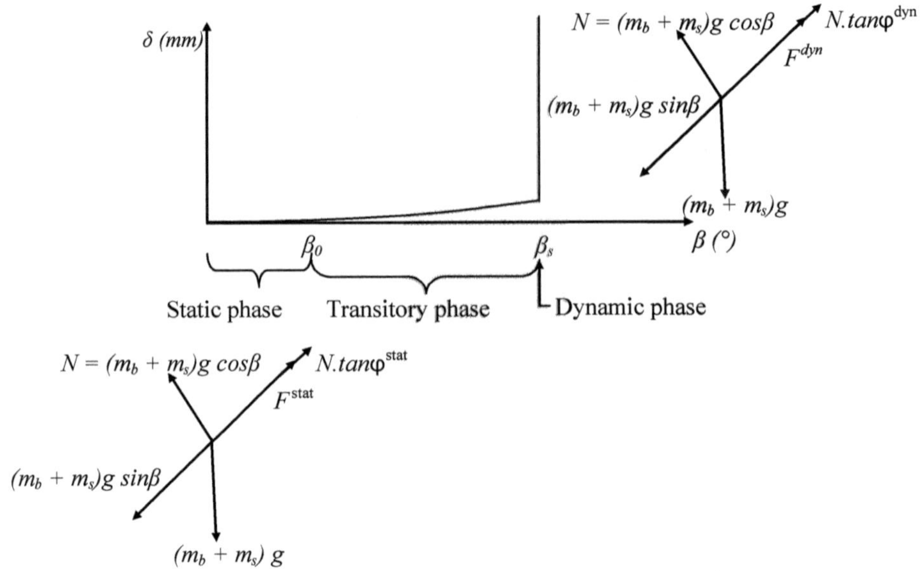

Fig. 13. Free-body diagram of 'Tension Procedure' of the inclined plane test

Table 5 presents the interface friction angles values calculated for the four tested geosynthetics of reinforcement using the following equations:

$$tan\,\varphi^{stat} = \frac{(m_b + m_s) \cdot g \cdot sin\,\beta_0 - F^{stat}}{m_s \cdot g \cdot cos\,\beta_0} \quad (5)$$

$$tan\,\varphi^{dyn} = \frac{(m_b + m_s) \cdot g \cdot sin\,\beta_s - F^{dyn}}{m_s \cdot g \cdot cos\,\beta_s} \quad (6)$$

Rzepecki et al. (2013) performed inclined plane tests on the same geosynthetic products and used the same inclined plane device under two different initial normal stress of 2 kPa and 2.2 kPa with the aim to compare the performance of the different geosynthetic of reinforcement.

Table 5. Soil-geosynthetic interface friction angles calculated in accordance with the "Tension Procedure", the "Static – Dynamic Procedure" and the standard procedure.

«Tension procedure»			«Static- dynamic procedure»		Standard procedure
Geosynthetic of reinforcement	φ^{stat} (°)	φ^{dyn} (°)	φ^{stat} (°)	φ^{dyn} (°)	φ^{stand}
Gtr_1	19	18	20	23,6	26,5
Gtr_2	22	21	23	23,6	25
$[Gtr_1\text{-}Gmat_1]$	18	16	19	28,5	32
$[Gtr_2\text{-}Gmat_2]$	22.5	21.5	23	29,6	33

Test results were interpreted on the “Static- Dynamic Procedure” of Gourc and Reyes-Ramirez (2004) and allowed the choice of the geocomposite [Gtr_2-$Gmat_2$] as the best geosynthetic of reinforcement among the tested products, similarly to this study results.

The major difference between the “Tension Procedure” and the “static – dynamic procedure” is that the later interprets the gradual sliding by an increase of interface friction with the movement $\left(\varphi^{dyn} > \varphi^{stat}\right)$ as shown on the Table 5.

On the other hand, the «tension procedure» considers the mechanical characteristics of tested geosynthetic which allows a more comprehensive interpretation of the geosynhetic behavior. The diagram geosynthetic tension (F) versus displacement (δ) provides additional information required to correctly understand the sliding mechanism and to assess correctly the interface friction angle.

Best geosynthetic of reinforcement are those mobilizing more the static friction to tension thanks to a better surface roughness and without elongation thanks to a high tensile modulus (J). In this case, more extended static phase is recorded and higher interface friction angle φ^{stat} can be calculated since the used value of β_0 is high.

If geosynthetic deformation is inevitable because of an excessive static friction mobilization under high initial normal stress and good surface roughness conditions, the best geosynthetic of reinforcement is that allowing more elongation allowing a gradual sliding of the retaining soil box on the plane. The relatively extended transitory phase prolongs the time before the inevitable non-stabilized sliding.

6 Conclusions

Inclined plane tests were conducted on two soil-geotextile and two soil-geocomposite interfaces.

Results were interpreted according to the standard method (EN ISO 12957-2), the “static- dynamic procedure” of Gourc and Reyes-Ramirez (2004) and a new procedure which considers the geosynthetic tension when calculating the interface friction angles.

The main advantages of the proposed “tension method” are:

- Provides much suitable information of the geosynthetic behavior during tests and its response in regard to the interface shear strength.
- The calculated interface friction angles seem to be more reliable than those calculated according to the standard and the “static- dynamic procedure” of Gourc and Reyes-Ramirez (2004).

References

Briançon, L., Girard, H., Gourc, J.P., Poulain, D.: Justification d’une modification de la norme relative au frottement des interfaces géosynthétiques au plan incliné. In: Proceedings Rencontres Géosynthétiques 2011, pp. 219–228 (2011)

Carbone, L.: Interface behavior of geosynthetics in landfill cover systems under static and seismic loading conditions. these de doctorat LTHE Grenoble, 245 p. (2013)

EN ISO 12957-2: Geosynthetics – determination of friction characteristics, Part 2: Inclined plane test. European committee for standardization, Brussels (2005)

Gourc, J.P., Reyes Ramirez, R.: Dynamics-based interpretation of the interface friction test at the inclined plane. Geosynth. Int. **11**(6), 439–454 (2004)

Gourc, J.P., Pitanga, H.N., Reyes-Ramirez, R.: Attempt of interpreting comprehensively the inclined plane test for geosynthetic interfaces. In: Proceedings of the First American Geosynthetic Conference and Exhibition, Cancun, Mexico, March 2008, 8 p. (2008)

Lopes, M.L., Ferreira, F., Carneiro, J.R., Vieira, C.S.: Soil geosynthetic inclined plane shear behavior, influence of soil moisture content and geosynthetic type. Int. J. Geotech. Eng. **8**(3), 335–342 (2014)

Rzepecki, D., Gourc, J.P., Dembicki, E.: Frictional characteristics of geosynthetic – geosynthetic and geosynthetic – soil interfaces determined by the inclined plane apparatus. Studia Geotechnica et Mechanica **XXXV**(4), 23–45 (2013)

Stoltz, G., Poulain, D., Loheas, E., Gallo, R., Touze Foltz, N.: Procédure d'essai au plan incline pour determiner les propriétés de frottement résiduel aux interfaces géosynthétiques. In: Proceedings Journées Nationales de Géotechnique et de Géologie de l'Ingénieur JNGG2012, Bordeaux, 4–6 juillet 2012, 9 p. (2012)

Stress-Strain Behaviour of Sand with Disc Plate Shaped Reinforcement

J.N. Jha[1(✉)], S.K. Shukla[2], A.K. Choudhary[3], K.S. Gill[1], and B.P. Verma[4]

[1] Guru Nanak Dev Engineering College, Ludhiana, Punjab, India
jagadanand@gmail.com
[2] Edith Cowan University, Perth, Australia
[3] National Institute of Technology, Jamshedpur, Jharkhand, India
[4] RVS College of Engineering and Technology, Jamshedpur, Jharkhand, India

Abstract. In the present study, the laboratory triaxial compression tests were carried out on soil specimens reinforced with steel and aluminium solid plates in horizontal layers. The percentages of reinforcement used were 5%, 4%, 2%, 1% and 0.5%. The solid plates were placed horizontally in five layers in all the tests. Again the triaxial compression tests were repeated by using the perforated circular aluminium plates as horizontal reinforcement instead of solid plate, but the quantity of reinforcements was kept the same as in the previous case. The diameter of the plate in all the cases was 25 mm. To alter the percentages of reinforcement, thickness of layers were varied in each case but thickness in all the five different layers were kept the same. The results show that improvement in strength of soil was not proportional to the increase in the percentage of reinforcement and residual strength ratio was also found to be less. It was also observed that there was an increase in the tangent modulus with increase in percentage of reinforcement at higher confining pressure when aluminium was used as a disc shaped plate reinforcement.

1 Introduction

The beneficial effect of using different reinforcing material largely depends on the form in which it is used as reinforcement. When exactly the same quantities of reinforcing material are used in different form like planer layers or discrete fibers, the strength improvement are different for different forms. Again if the same quantities of reinforcement having the same shape are used for different material, strength improvements will be different. This difference in strengths achieved is mainly due to difference in mechanism of failure in the soil reinforced with different form/material. Horizontal layers improve the strength mainly by friction and interlocking between soil and reinforcements whereas the randomly oriented fibers improve the strength by friction and coiling around the soil particles. Much research has been carried out to understand the beneficial effects of planar form of geosynthetic reinforcement in sand and randomly oriented discrete geofibers to reinforce the sand (Haeri et al. 2000; Venkatappa Rao et al. 2005; Madhavi Latha and Murthy 2007; Choudhary et al. 2010; Jha et al. 2014, 2017; Butt et al. 2016). Limited studies are also available on sand reinforced with

S.K. Shukla and E. Guler (eds.), *Advances in Reinforced Soil Structures*,
Sustainable Civil Infrastructures, DOI 10.1007/978-3-319-63570-5_3

galvanized iron sheet and hard plastic sheets (Verma and Char 1978; Zhang et al. 2006). This paper reports the relative efficiency of solid circular plates made of steel and aluminium when the reinforcement quantity remains the same by conducting systematic series of triaxial compression tests on reinforced sand with these two forms of circular plates reinforcements. Again the triaxial compression tests were repeated by using perforated circular aluminium plate as horizontal reinforcement instead of solid plate, but the quantity of reinforcements was kept the same as in the previous case. The percentages of reinforcement used for both series of tests were 5%, 4%, 2%, 1% and 0.5%. The results were analysed to compare the form of reinforcement and to study the effect of reinforcement type on the strength properties of sand.

2 Materials and Methods

2.1 Steel and Aluminium

Aluminium sheet having a thickness of 0.45 mm and mild steel sheet having a thickness of 0.16 mm procured from the local market were used as the reinforcing material. Circular plates were cut from these sheets. The stress-strain curve shown in Fig. 1 was used to determine the modulus of elasticity of the materials.

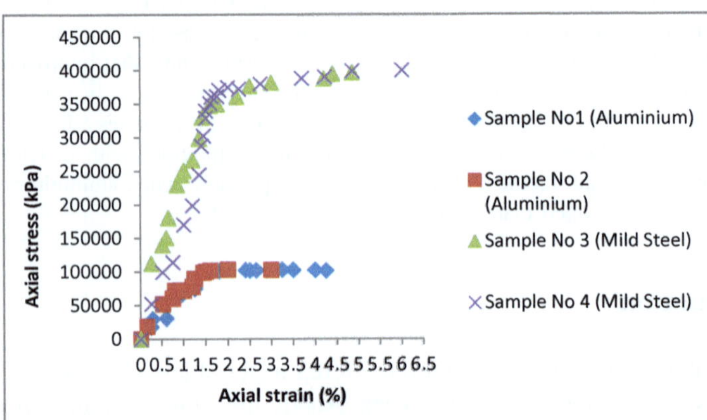

Fig. 1. Stress strain behaviour of reinforcing material

2.2 Sand

Standard Ennore sand was used in the present investigation. The average particle diameter (D_{50}) was 0.68 mm. The uniformity coefficient (C_u) and effective size (D_{10}) of the sand used were 1.408 and 0.49 mm, respectively. Grain-size distribution of the Ennore sand is given in Fig. 2. The friction coefficients between the Ennore sand and the two materials (mild steel and aluminium) obtained from direct shear tests were 0.445 and 0.404, respectively.

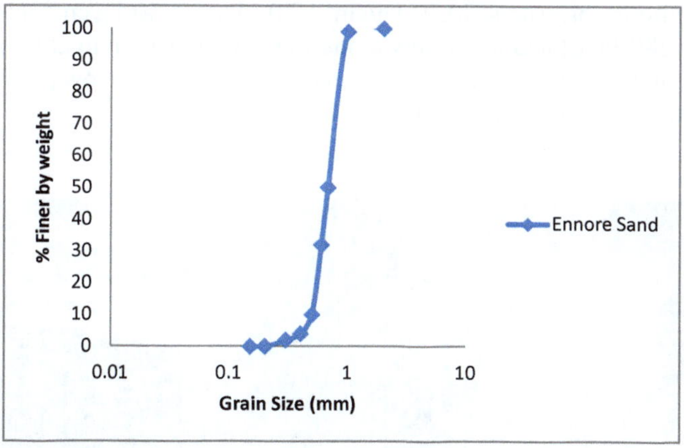

Fig. 2. Grain size distribution

2.3 Triaxial Compression Test

Triaxial compression tests for both series were performed on air-dried sand. In the first series, experiments were conducted using solid circular plates (mild steel and aluminium) whereas in the second series, perforated plates (aluminium) were used. The plates were kept horizontally in five layers in all the tests for both series. The diameter of the plates in all the cases was 25 mm. To alter the percentage of reinforcement, thickness of the each layer was varied but the thickness in all the five different layers was kept constant. Five plates of 25 mm dia. having 0.16 mm thickness gave 0.5% of reinforcement by volume for mild steel plate. 1% of reinforcement was obtained by 10 plates, 2% by 20 plates, 3% by 30 plates, 4% by 40 plates and 5% by 50 plates. Since each time 5 layers of reinforcements were to be used, the plates were fixed in groups of 2, 4, 6, 8 and 10 by araldite adhesive. This was done to avoid the sliding of plates among themselves when subjected to external stresses. The required thickness for aluminium plates were also computed for different percentages of reinforcement and the number of layers of reinforcement in this case was again kept at five. All tests were conducted on samples of 38 mm diameter with an aspect ratio 2. The test specimens were prepared by a procedure similar to that adopted for preparing specimens of sand for conventional unconsolidated undrained triaxial tests. The sand was filled in the split mould in six layers and each time 50 tampings were given with 6.35 mm diameter glass rod. To get the fairly uniform density for all the samples, this tamping method was used several times for sample preparation before running the tests. The weight of the total amount of sand was used to determine the density. The density for different sample was 1.5 g/cm^3 and this was maintained constant for all cases. The amount of sand to fill one sixth of the split mould was measured in a container and the same was used to fill the sand in the sampler. This container was used to place the sand in six layers. After putting the first layer required number of tamping was given by the glass rod. The plate reinforcement was placed horizontally on the sand layers and the next layer of sand was poured. The same procedure was repeated till the split mould was filled in six layers. All tests were conducted at a strain rate of 1.25 mm/min for three different confining pressures [0.70 kg/cm^2 (70 kPa), 1.40 kg/cm^2 (140 kPa) and 2.80 kg/cm^2 (280 kPa)] and the observations were continued up to at least 20% strain. The experimental set up showing the experiment in progress is shown in Fig. 3 and the variables of study are given in Table 1.

Fig. 3. Experimental set up

Table 1. Variables of study

Reinforcement type	Percentage of reinforcement	Confining pressure (kPa)
Solid circular plate - A (mild steel)	0.5, 1, 2, 3, 4, 5	70, 140, 280
Solid circular plate - B (aluminium)	0.5, 1, 2, 4, 5	70, 140, 280
Perforated circular plate - C (aluminium)	1, 2, 4	70, 140, 280

3 Results and Discussion

The stress strain behaviour of the unreinforced sand at confining pressures of 70, 140 and 280 kPa is shown in Fig. 4. The shear strength parameters of unreinforced sand are $c = 0$ and $ø = 36°$. Typical stress-strain relationships of sand reinforced with an equal amount of reinforcements for the three different forms of reinforcements (A- Solid Disc Circular Plate Mild Steel, B- Solid Disc Circular Aluminium Plate, C-Perforated Disc Circular Aluminium Plate) are shown in Fig. 5. Quantity of reinforcement considered in all the three cases is 4%.

Comparing the stress strain behaviour of reinforced case (Fig. 5) with unreinforced case (Fig. 4), it can be observed that all reinforced specimens, exhibited improved stress–strain response in terms of increase in peak deviator stress and increased failure strains. Some typical peak deviator stress value at different percentage of strain has

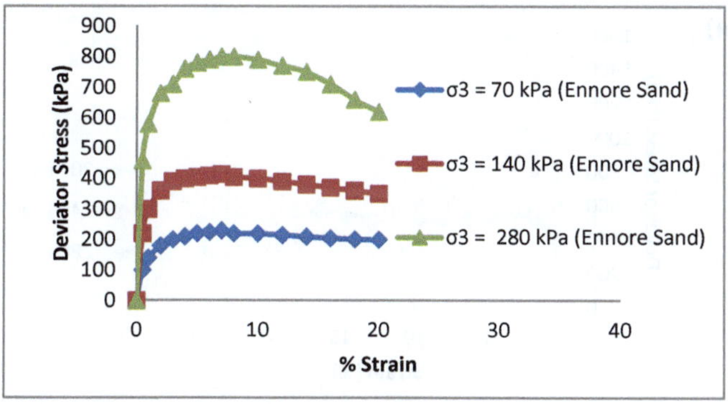

Fig. 4. Stress-strain behaviour of unreinforced sand

been given in Table 2 and it can be observed that at a given confining pressure, the peak deviator stress and corresponding strain for unreinforced case was always less than their corresponding value for the reinforced case.

Table 3 shows the variation of strain at failure, peak deviator stress and the strength ratio for sand reinforced with different forms of reinforcement at a given percentage of reinforcement with increasing confining pressure. The strength ratio is defined as the peak deviator stress of reinforced specimen to the peak deviator stress of unreinforced specimen. The peak deviator stress for unreinforced sand at 70 kN/m^2, 140 kN/m^2 and 280 kN/m^2 are 205 kN/m^2, 405 kN/m^2, 790 kN/m^2 respectively. Now comparing this value with reinforced case, It can be observed from Table 3 that the improvement in peak deviator stress ranges from 2.75–3.68 for A (Solid Disc Circular Plate Mild Steel), 2.46–3.55, for B (Solid Disc Circular Aluminium Plate) and 2.42–2.98 for C (Perforated Disc Circular Aluminium Plate) respectively. Similarly the strength ratio ranges from 1.44–2.25 for A (Solid Disc Circular Plate Mild Steel), 1.84–2.60, for B (Solid Disc Circular Aluminium Plate) and 1.85–2.77 for C (Perforated Disc Circular Aluminium Plate), respectively. Result of strength ratio shows that aluminium as reinforcement is more effective in improving the strength ratio as compare to mild steel, despite the fact that it has high tensile strength than aluminium. Hence it can be concluded that improvement is not dependent on tensile strength alone, but other factors like ductility and confining stress also plays a significant role in enhancing the strength ratio and peak strength.

The shear strength parameters have been determined from the *p*–*q* diagram at failure for sand reinforced with different form of reinforcements. The values of *c* and *ø* obtained from *p*–*q* plots for various forms and different types of reinforcing materials are summarized in Table 4. As observed from the table, the shear strength parameters varied for different form and types of reinforcing materials. The difference in the friction angle is marginal. But the *c* value is observed to be quite sensitive to the reinforcement form and type. As can be seen from the table that when the same amount of reinforcement is used, the perforated aluminium is found to give the value of cohesion '*c*' in the range of 1.4–2.4 times more than that of mild steel which has comparatively a very high tensile strength though the increase in friction angle is

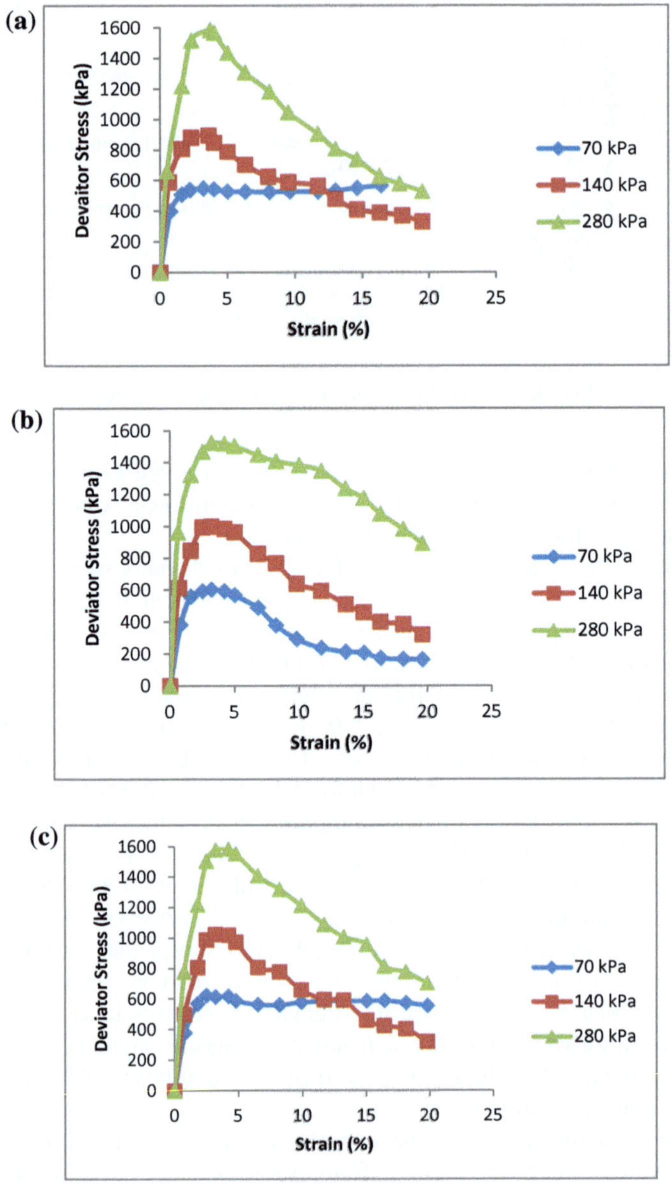

Fig. 5. (a) - Stress-strain behaviour of reinforced sand (A-Solid Disc Circular Plate Mild Steel). (b) - Stress-strain behaviour of reinforced sand (B-Solid Disc Circular Aluminium Plate). (c) - Stress-strain behaviour of reinforced sand (C-Perforated Disc Circular Aluminium Plate)

marginal. The possible reason of the improved performance of the perforated aluminium inclusion is related to the presence of the perforations which allow soil-to-soil frictional resistance.

Table 2. Peak deviator stress (kN/m^2) at different %age of strain for different type of reinforcement

Confining pressure (kN/m^2)	Unreinforced		Strain (%) - 8			Strain (%) - 10			Strain (%) - 12		
			Type of reinforcement			Type of reinforcement			Type of reinforcement		
	Strain (%)	Peak stress	A	B	C	A	B	C	A	B	C
70	6	205	527	380	562	530	295	580	530	240	585
140	8	405	625	770	780	590	640	660	565	595	595
280	6	790	1185	1410	1320	1050	1385	1215	910	1350	1090

A (Solid Disc Circular Plate Mild Steel).
B (Solid Disc Circular Aluminium Plate).
C (Perforated Disc Circular Aluminium Plate).

Table 3. Strength characteristics of reinforced sand

Percentage of reinforcement	σ_3 kPa	Percentage strain at failure			Peak stress (σ_1–σ_3) kPa			Strength ratio		
		Reinforcement type			Reinforcement type			Reinforcement type		
		A	B	C	A	B	C	A	B	C
1.0	70	4.0	3.5	3.4	410	620	650	1.73	2.60	2.77
	140	3.6	3.6	4.8	605	900	1050	1.44	2.14	2.54
	280	5.5	5.0	3.5	1130	1680	1570	1.40	2.08	1.85
2.0	70	3.4	3.4	2.5	535	420	550	2.25	1.76	2.31
	140	4.0	3.6	3.4	815	820	1040	1.94	1.95	2.47
	280	5.5	3.8	5.0	1414	1490	1640	1.75	1.84	2.02
4.0	70	2.4	3.4	3.2	430	615	650	1.81	2.58	2.73
	140	3.2	3.3	4.0	895	1005	1050	2.13	2.39	2.50
	280	3.3	4.1	3.9	1580	1510	1570	1.96	1.87	1.94

A-Solid Disc Circular Plate (Mild Steel), B-Solid Disc Circular Plate (Aluminium), C-Perforated Disc Circular Plate (Aluminium), σ_3 – Confining Pressure.

Table 4. Shear strength parameters of reinforced sand

Percentage of reinforcement	c (kPa)			$\o^\circ$		
	Reinforcement type			Reinforcement type		
	A	B	C	A	B	C
0.5	40	60		39	44	
1.0	40	60	96	39	45	42
2.0	45	60	65	43	45	45
3.0	45			44		
4.0	55	80	80	44	42	44
5.0	27	80		47	44	

Initial tangent modulus is the slope of the line which is tangent to the stress-strain curve at zero load, and it gives an idea of initial stiffness and the elastic range of the material. It can be observed from Table 5 that when the same amount of reinforcement is used, the initial tangent modulus increases with an increase in confining pressure for all the type and form of reinforcement. Secant modulus is the slope of the line which is joined by any point on the stress strain curve with the origin was determined at the maximum deviator stress. For a given amount of reinforcement, the secant modulus also increases with an increase in confining pressure. With few exceptions, it can be observed from the result of Table 5 that the value of moduli depends on amount of reinforcement/confining stress or both and aluminium reinforcement was found to be far more effective than the mild steel despite having low tensile strength. The possible reason for such result is due to the high ductility of aluminium.

Table 5. Elastic moduli of reinforced sand

Percentage of reinforcement	σ_3 kPa	Initial tangent modulus (kPa)			Secant modulus at failure (kPa)		
		Reinforcement type			Reinforcement type		
		A	B	C	A	B	C
1.0	70	45000	51250	50000	10250	17714	19412
	140	46250	75000	58330	16806	25000	22292
	280	55000	88333	100000	20545	33600	42857
2.0	70	40000	38120	44440	15735	12353	22000
	140	70000	78570	75000	20375	22778	30588
	280	52220	108333	80000	25709	31042	32800
4.0	70	50000	50000	42857	17917	18088	20313
	140	72000	80000	58570	27969	30455	26250
	280	95000	120000	83330	47879	36829	40256

The ratio of residual stress to the peak deviator stress is the residual strength ratio and has been tabulated in Table 6. Residual strength ratio ranges from 0.32–0.81 for A (Solid Disc Circular Plate Mild Steel), 0.29–0.71 for B (Solid Disc Circular Aluminium Plate) and 0.24–0.75 for C (Perforated Disc Circular Aluminium Plate) respectively. It also depends on confining pressure and percentage of reinforcement. Since surface of mild steel is more rough as compare to aluminium, it is observed that solid circular plate mild steel is more effective in retaining the residual strength than aluminium. The loss in strength is also partially due to destruction of frictional bonds and reorientation of particles. Again it is observed from Table 6 that at a given amount of reinforcement, the equivalent confining stress increase ($\Delta\sigma_3$) was not found to be effective at low confining pressure but starts increasing with increases in confinement which increases with increase confining stress.

Table 6. Residual strength ratio and equivalent confining pressure increase for reinforced sand

Percentage of reinforcement	Confining pressure σ_3 kPa	Residual strength ratio			Equivalent confining pressure increase $\Delta\sigma_3$ kPa		
		Reinforcement type			Reinforcement type		
		A	B	C	A	B	C
1.0	70	0.51	0.29	0.24	0.68	0.98	1.07
	140	0.45	0.49	0.43	1.38	1.25	1.68
	280	0.54	0.63	0.75	2.30	2.25	1.79
2.0	70	0.36	0.50	0.62	0.48	0.46	0.80
	140	0.36	0.70	0.50	1.23	1.04	1.61
	280	0.81	0.56	0.49	2.00	1.77	2.15
4.0	70	0.64	0.32	0.56	0.67	0.96	0.95
	140	0.65	0.32	0.50	1.38	1.52	1.68
	280	0.32	0.56	0.40	2.30	1.82	1.97

4 Conclusions

Based on the experimental results, the following general conclusions can be drawn.

1. All reinforced specimens exhibited improved stress-strain response compared to unreinforced sand at all confining pressures and forms in terms of improved peak deviator stress and increased strength ratio.
2. Improvement in peak deviator stress and strength ratio is not proportional to the increase in reinforcement amount. These results show that despite being low tensile strength of aluminium, its performance in improving the peak stress and strength ratio is comparable to the performance of mild steel which has comparatively high tensile strength.
3. The cohesion value is observed to be quite sensitive to the reinforcement form and type. When the same amount of reinforcement is used, the perforated aluminium is found to give the value of cohesion '*c*' in the range of 1.4–2.4 times more than that of mild steel, but the increase in friction angle is marginal.
4. For a given amount of reinforcement, the initial tangent modulus and secant modulus increase with an increase in confining pressure. Despite being low tensile strength, aluminium reinforcement was found to be far more effective than the mild steel in both the cases.
5. At a given amount of reinforcement, the equivalent confining stress increase ($\Delta\sigma_3$) was not found to be effective at low confining pressure but starts increasing with increases in confining pressure.
6. Residual strength ratio ranges from 0.32 to 0.81 depending on the type and form of reinforcement. Solid circular plate mild steel was found to be most effective in retaining the residual strength among the type of reinforcement used.

References

Haeri, S.M., Noorzad, R., Oskoorouchi, A.M.: Effect of geotextile reinforcement on the mechanical behavior of sand. Geotext. Geomembr. **18**, 385–402 (2000)

Venkatappa Rao, G., Dutta, R.K., Ujwala, D.: Strength characteristics of sand reinforced with coir fibers and coir geotextiles. Electron. J. Geotech. Eng. USA 10/G (2005). http://www.ejge.com

Madhavi Latha, G., Murthy, V.S.: Effect of reinforcement form on the behaviour of geosynthetic reinforced sand. Geotext. Geomembr. **25**, 23–32 (2007)

Choudhary, A.K., Jha, J.N., Gill, K.S.: A study on CBR behaviour of waste plastic strip reinforced soil. Emir. J. Eng. Res. **15**(1), 51–57 (2010)

Jha, J.N., Choudhary, A.K., Gill, K.S., Shukla, S.K.: Behaviour of plastic waste fibre- reinforced industrial wastes in pavement applications. Int. J. Geotech. Eng. **8**(3), 277–286 (2014)

Jha, J.N., Shukla, S.K., Choudhary, A.K., Gill, K.S.: Triaxial behaviour of steel reinforced soil. In: ASCE Conference: Geo Frontier (2017, paper submitted for acceptance)

Butt, W.A., Mir, B.A., Jha, J.N.: Strength behaviour of clayey soil reinforced with human hair as natural fiber. Int. J. Geotech. Geol. Eng. **34**(1), 411–417 (2016)

Verma, B.P., Char, A.N.R.: Triaxial tests on reinforced sand. In: International Symposium on Soil Reinforcing and Stabilising Techniques, Sydney, Australia, pp. 29–39 (1978)

Zhang, M.X., Javadi, A.A., Min, X.: Triaxial tests of sand reinforced with 3D inclusions. Geotext. Geomembr. **24**, 201–209 (2006)

Swelling and Shrinkage Behaviour of Expansive Soil Blended with Lime and Fibres

Mayakrishnan Muthukumar[1(✉)], S.K. Sekar[2], and Sanjay Kumar Shukla[3]

[1] Department of Structural and Geotechnical Engineering, School of Civil and Chemical Engineering, VIT University, Vellore, India
mmuthukumar@vit.ac.in

[2] School of Civil and Chemical Engineering, VIT University, Vellore, India
sksekar@vit.ac.in

[3] Discipline of Civil and Environmental Engineering, School of Engineering, Edith Cowan University, Perth, Australia
s.shukla@ecu.edu.au

Abstract. Expansive soils are considered to be highly problematic because of their capacity to significant volume change. They swell during the rainy season as they absorb water and shrink when water evaporates from them during the summer season. Because of this dual swell-shrink behaviour, an expansive soil causes severe distress to many civil engineering structures. Several mitigating techniques are adopted to counteract the problems posed by the expansive soils, either by modifying the properties of the soil by adopting stabilization techniques using lime, cement, fly ash, calcium chloride etc. or by adopting special foundation technique such as construction of belled piers, under-reamed piles, etc. In recent years polymeric fibres have also been used to stabilize the soil as well as to improve the strength of the expansive soils. Hence in the present study lime and fibres have been used in different proportions to study the swelling and shrinkage behavior of expansive soils. Swell tests were performed by varying the fibre content and lime with expansive soils. Tests were also conducted by blending fibres and lime together with expansive soils. In a similar way, shrinkage tests were also performed for the various proportions. The test result show that swelling tends to decrease slightly with an increase in the fibre content, whereas shrinkage tends to decrease significantly upon addition of fibres. Both swelling and shrinkage tends to decrease significantly with increasing lime content. The optimum content of fibre was found to be 2%. So the expansive soil specimens blended with 2% fibres and with varying lime content was tested. It is found that blending 2% fibres and 15% lime together in expansive soils is considered to be more effective in controlling the swelling and shrinkage behaviour.

1 Introduction

The problems posed by expansive soils have been recorded worldwide. These expansive soils swell during the rainy season and shrinks during the summer seasons. This cyclic volume change behaviour of expansive soils causes severe damage to the

S.K. Shukla and E. Guler (eds.), *Advances in Reinforced Soil Structures*,
Sustainable Civil Infrastructures, DOI 10.1007/978-3-319-63570-5_4

lightly loaded structure founded on them (Chen 1988). To minimize the volume change attributes of these expansive soils, several mitigating techniques have been adopted. Stabilization of expansive soil with various additives, including lime, cement, calcium chloride and fly ash has shown promising results in heave reduction and improved strength characteristics (Shanker and Maruthi 1989; Cokca 2001; Sharma 1998).

Al-Rawas et al. (2002) studied the effect of lime, cement and artificial pozzolan and the combination of these three stabilizers at different proportions. Lime, cement and sarooj, an artificial pozzolana produced by burning calcining clay were mixed at different dosages by dry weight of soil. It was observed that swell percentage and swell pressure reduced to zero at 6% of lime. Rice husk ash stabilized with lime or cement was used as cushion between the expansive soil and the foundation to counteract the effect of heaving of expansive soils (Sivapullaiah et al. 2004; Sharma et al. 2008).

Fiber reinforcement of expansive soils is also found to be successful in reducing volume changes and increasing the shear strength of expansive soils. The effect of discrete and randomly oriented polypropylene fibre reinforcement on strength and volume change behaviour of expansive soils was studied by Puppala and Musenda (2000). Fiber reinforced clayey samples were prepared by varying the fibre percentage as 0%, 0.3%, 0.6% and 0.9% by dry weight of soil for the both types of fibers. Test results from their studies showed that the fiber reinforcement enhanced strength and reduces volumetric shrinkage and swelling pressure. Fiber reinforcement also decreases swell potential considerably.

Al-Akhras et al. (2008) carried out investigation on expansive soil with nylon fibres and natural fibres having different aspect ratios to study the influence of fibres on swelling properties of clayey soils. Four aspects ratios (*l/d*) of 25, 50, 75 and 100 and five different fibre contents of 1%, 2%, 3%, 4% and 5% were used in the study. Results revealed that both swelling pressure and swell potential reduced significantly with an increase in the fiber content. From their study it was also observed that natural fibres are more efficacious in controlling heave than nylon fibres. Further, a lower aspect ratio appeared to have a greater effect in reducing swelling pressure in both types of fibres.

The effect of polypropylene tape fibers on swelling behaviour of expansive soils was studied by Viswanatham et al. (2009). One-dimensional swell tests were conducted on remoulded expansive soils without reinforcement and with reinforcement. The percentage range of 0.25% and 0.5% were used with lengths of 30 mm, 60 mm and 90 mm. The study revealed that reduction in heave was proportional to fiber content and maximum heave was observed at a low aspect ratio for both the fibre contents of 0.25% and 0.5%. It was also observed that the length of the fibre was the key factor that influenced the reinforcing effect of fibre. Discrete and randomly distributed fibers were found efficacious in reducing heave.

Puppala (2001) carried out investigation on mixture of fiber and fly ash to stabilize expansive soils. This technique was also found effective in reducing plasticity and free swell characteristics. Kumar et al. (2007) studied the effect of fly ash, lime and polyester fibres on compaction and strength properties of expansive soils. Randomly oriented fibres were introduced in mixes at different percentages and observed that strength increased with increase in curing period.

From the literature review, it is observed that mostly studies are carried out only on reducing the swelling nature of the expansive soils, but there is limited research available on controlling both swelling and shrinkage of expansive soils. In the present study, the swelling and shrinkage behaviour of expansive soils have been studied by adding varying lime content, fibre content and blending both lime and fibres with expansive soils at different percentages.

2 Experimental Investigation

2.1 Test Materials

Bentonite, commercially available clay, was used for the present investigation. The bentonite used was sodium bentonite because it will undergo more volume change when compared to other types of bentonite. Various index properties of the soil were determined and presented in Table 1. Based on liquid limit and plasticity index of the soil, the soil was classified as CH according to Unified Soil Classification System (USCS). The free swell index of the soil was found to be 200%, which is considered to be for a highly swelling expansive soil.

Table 1. Properties of the soil used for this study

Soil properties	Value
Specific gravity	2.68
Liquid limit %	121
Plastic limit %	48
Shrinkage limit %	8
Plasticity index %	82
Free swell index (FSI) %	200
Classification according to USCS	CH

Polypropylene fibres have been used for the present study because of its several advantages like high strength, micro fine reinforcement, chemically inert, noncorrosive and available in varying length. For the present study, fibre length of 6 mm was used.

Commercially available hydrated lime $Ca(OH)_2$ was used for the present study, because of the difficulty in handling quick lime. Fibre has been denoted as ‘F’ and Lime has been denoted as ‘L’ in the present study.

2.2 Swelling Tests

Fibre content was varied as 1%, 2%, 4% and 6% in swelling tests. In another series of tests lime content was varied as 1%, 2% and 4%. The results of the swelling tests with varying fibre content indicated that, the swelling was less with 2%. Hence the effect of swelling behaviour of expansive soil with 2% fibre was studied by varying the lime content (1%, 2% and 4%). The oven dried expansive soil was mixed with initial water content of 10% and with a dry unit weight of 12 kN/m^3 in the mould of 10 cm diameter in three layers of thickness 25 mm. A dial gauge was fitted at the top of the soil layer.

After setting the dial gauge reading to zero, water was added continuously at the top of the clay bed and the swelling was monitoring continuously at various time intervals until an equilibrium heave was attained.

2.3 Shrinkage Limit Tests

The fibre was mixed thoroughly with expansive clays for various percentages. The fibre content was varied as 0, 0.5%, 0.75%, 1% and 2%. The mixture was placed in the shrinkage dish and the specimen was oven dried. Shrinkage limit was determined for varying lime content (2%, 5%, 7.5%, 10% and 15%). Shrinkage limit was also determined on clay blends with 2% fibre and with varying lime content as mentioned earlier.

3 Discussion of Test Results

3.1 Influence of Fibre and Lime on the Swelling Behavior of Expansive Soils

Figure 1 shows the variation of swelling (mm) with varying lime content (%). It is observed that for a given fibre content, the swelling tends to increase with an increase in the time period but the swelling tends to decrease with an increase in the fibre content up to 2%. This is mainly because the fibre-reinforced soils behave like a composite materials in which the fibre having relatively high strength offering more

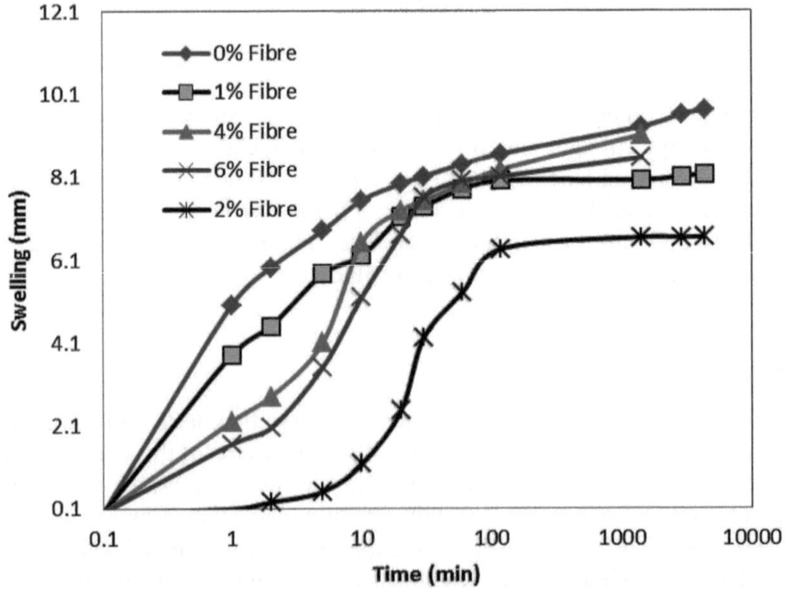

Fig. 1. Rate of heave for varying fibre content (%)

tensile resistance to soil against swelling. For 2% fibre content the initial swelling was less but beyond 2% fibre content swelling tends to increase. This is attributed due to the fact that beyond optimum fibre content further increase in the fibre content fails to offer good bonding between the soil and fibre which leads to an increase in heaving of expansive soil. It is found that the swelling potential of the expansive soil decreased from 21% to 15% with a fibre content of 2%.

Figure 2 shows the rate of swelling of expansive soil blended with varying lime content. The swelling tends to decrease with increase in the lime content. The swelling decreased from 9.5 mm to 2.5 mm, when the lime content was 4%, indicating 74% reduction in swelling. The reduction in swelling is attributed to the ions exchange process and pozzolanic reactions between the soil and lime, resulting in reduced amount of swelling.

Figure 3 shows the comparison of variation of swelling for a lime content of 4% and for a fibre content of 2%. It can be seen from the figure that the initial swelling is controlled effectively by fibres. After certain period of time as the water starts permeating into the soil, the soil get fully saturated and hence the heave tends to increase. In the case of expansive soil blended with lime, the initial swelling is more because of the slow process in the pozzolanic reactions (See Fig. 3). Hence further study has been carried out to reduce the initial swelling of clay lime blends by adding 2% fibre content, so that the total or final swelling will be further reduced.

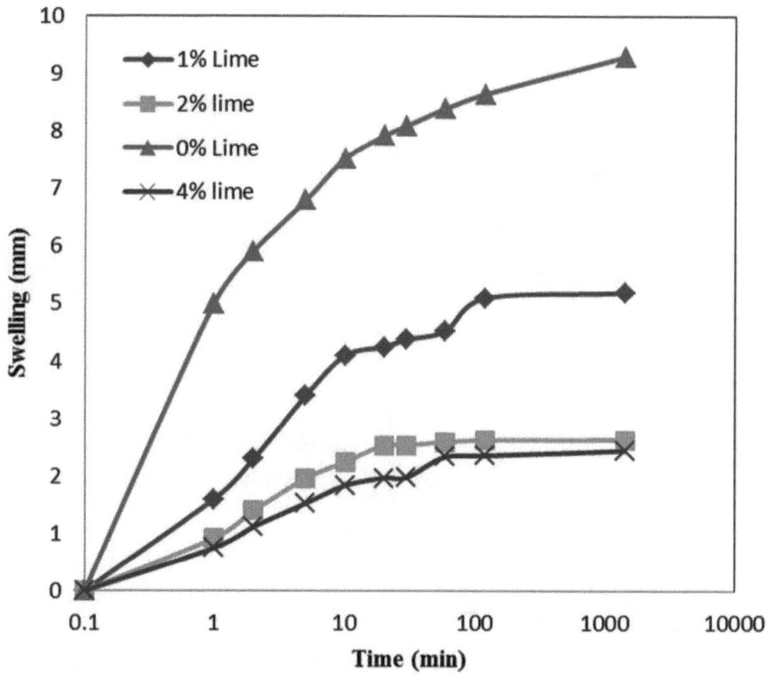

Fig. 2. Rate of heave for varying lime content (%)

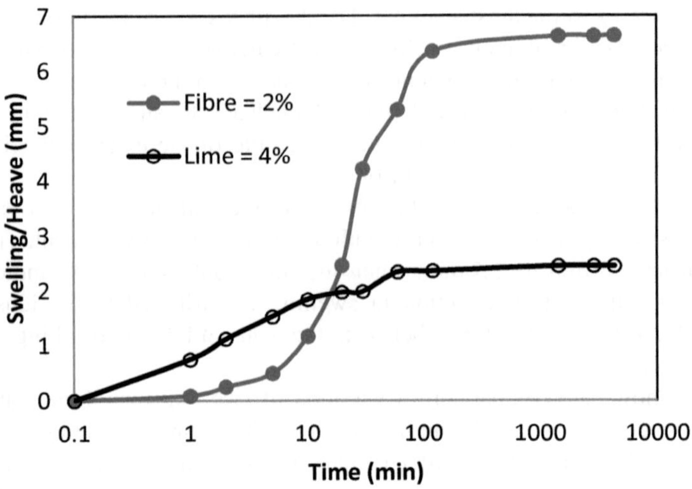

Fig. 3. Rate of heave with time

Figure 4 shows by comparison the variation of swelling with lime and the variation of swelling with lime plus 2% fibre content. It is clearly seen that swelling decreased significantly when 2% of fibre was added to lime-clay blends, indicating a potential decrease in the initial swelling and the further swelling is effectively controlled by lime

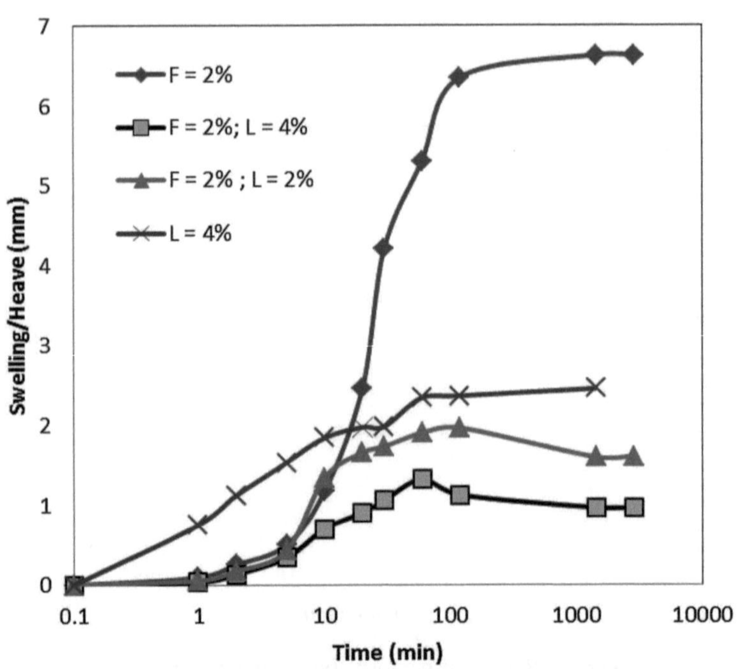

Fig. 4. Rate of heave with time

due to the pozzolanic reaction, resulted in lesser amount of heave when compared to the heave of expansive clay blends with lime.

3.2 Influence of Fibre and Lime on the Shrinkage Behavior of Expansive Soils

Figure 5 shows the effect of fibre content on the shrinkage behavior of expansive soils. It is observed that the shrinkage limit of the expansive soil tends to increase with an increase in the fibre content. The shrinkage limit is an indication of probable volume change of expansive soils. Lesser the shrinkage limit, higher the probable change in the volume of the soil (Holtz and Gibbs 1956). The shrinkage limit of the expansive soil is found to be 8% which is considered to be undergoing severe volume according to Holtz and Gibbs. When the fibre content of 0.5% was added, the shrinkage limit is found to be 38.8% indicating less volume change. The shrinkage limit tends to increase further upon addition of fibre and it is observed that there is no significant increase in shrinkage limit beyond 2% (See Fig. 5). This shows that the fibre reinforcement is proved to be effective not only in controlling the swelling but also the shrinkage of the expansive soil as well.

Figure 6 shows the variation of shrinkage limit for lime-clay blends of different proportions. In lime clay blends, the shrinkage limit also tends to increase with an increase in the lime content. The shrinkage limit is found to be 16.8% for a lime content of 2%, and for a lime content of 15% the shrinkage limit is found to be 36%. Beyond 15% there is no significant increase in the shrinkage limit (Fig. 6). The value of shrinkage limit is relatively less when compared to the fibre-clay blends with a maximum value of shrinkage limit of 58.6%.

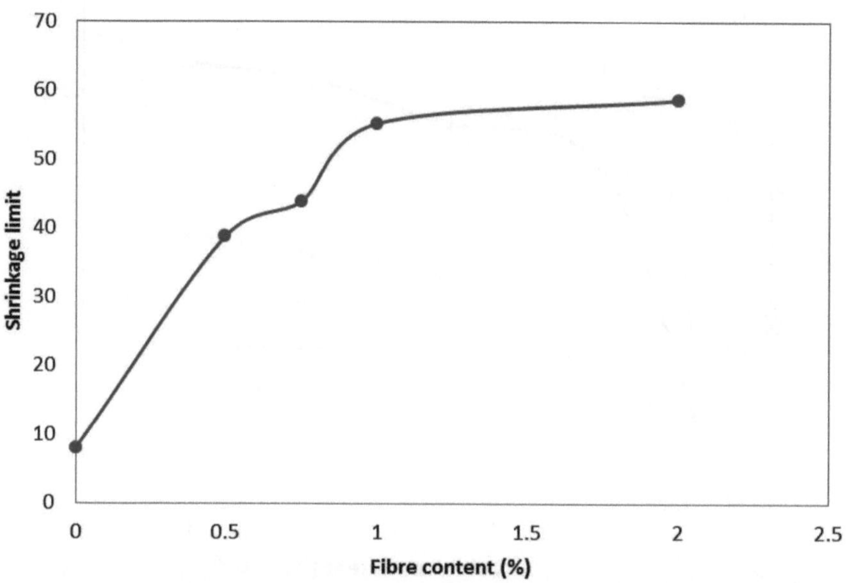

Fig. 5. Variation of shrinkage limit with fibre content

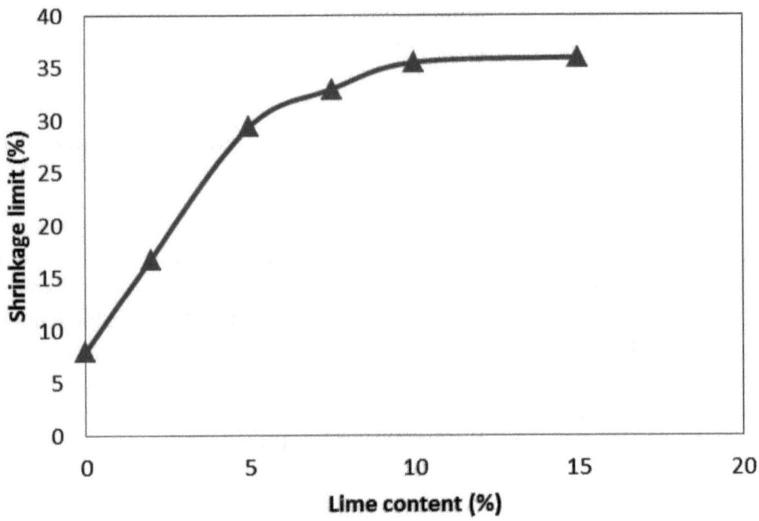

Fig. 6. Variation of shrinkage limit with lime content

Figure 7 shows by comparison, the variation of shrinkage limit for soil blended with lime, fibre and lime with 2% fibre. The shrinkage limit tends to increase significantly upon addition of fibre to any lime content. The shrinkage limit also increases marginally when compared to fibre-clay blends. From the shrinkage limit studies, it is

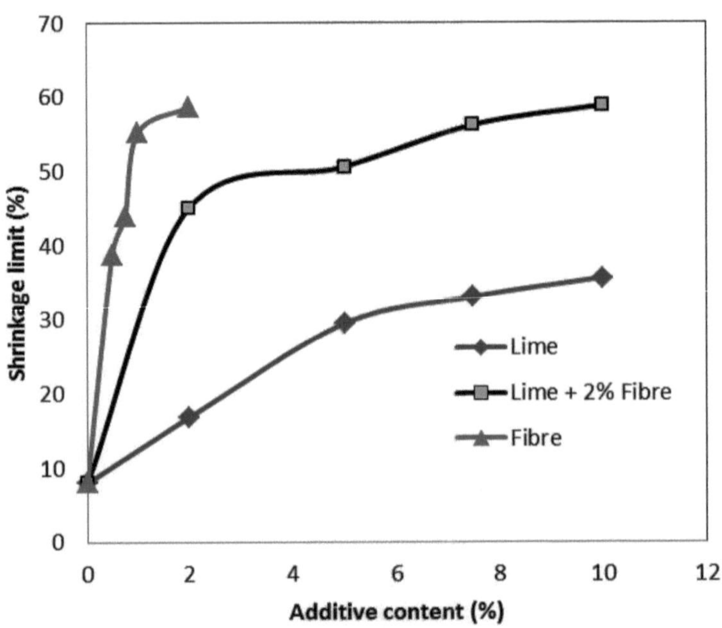

Fig. 7. Variation of shrinkage limit with additives (%)

observed that shrinkage limit is high when expansive soils are blended with fibre and it is true for any amount of additive, on the other hand the shrinkage cracks are more in the case of fibre-clay blends when compared to lime clay blends. The combination of fibre content of 2% and for a lime content of 15%, the shrinkage limit is found to be high (58.63%) and the development of shrinkage cracks also arrested effectively.

4 Conclusions

The following are the main conclusions arrived from the present study:

(1) Addition of fibres up to 2% to the expansive soil causes the heave to decrease. This is mainly because the fibre-reinforced soils behave like a composite materials in which the fibre having relatively high strength offering more tensile resistance to soil against swelling. The swelling potential of the expansive soil decreases from 21% to 15% with a fibre content of 2%.

(2) The swelling tends to decrease with an increase in the lime content. The swelling reduced 74%, when lime of 4% was added to expansive soils. Beyond 4% there is no further reduction in heave because of the initial swelling.

(3) Swelling decreased significantly when 2% of fibre is added to lime-clay blends, indicating a potential decrease in the initial swelling and the further swelling is effectively controlled by lime due to the pozzolanic reaction, resulting in lesser amount of heave when compared to the heave of expansive clay blends with lime.

(4) The maximum value of shrinkage limit is 36% for an optimum lime content, which is relatively less when compared to the fibre-clay blends with a maximum value of shrinkage limit of 58.6%.

(5) With the combination of fibre content of 2% and for a lime content of 15%, the shrinkage limit is found to be high (58.63%) and the development of shrinkage cracks are seen to be arrested effectively. Hence addition of fibre content of 2% and lime content of 15% can control both swelling and shrinkage effectively.

References

Al-Akhras, N.M., Altom, M.F., Al-Akhras, K.M., Malkawi, A.I.H.: Influence of fibres on swelling properties of clayey soil. Geosynth. Int. **15**(4), 304–309 (2008)

Al-Rawas, A.A., Taha, R., Nelson, J.D., Al-Shab, T.B., Al-Siyabi, H.: A comparative evaluation of various additives used in the stabilization of expansive soils. Geotech. Test. J. **25**(2), 199–209 (2002)

Chen, F.H.: Foundations on Expansive Soils, 2nd edn. Elsevier Scientific Publishing Co., Amsterdam (1988)

Cokca, E.: Use of class C fly ash for the stabilization of an expansive soil. J. Geotech. Geoenviron. Eng. **127**(7), 568–573 (2001)

Holtz, W.G., Gibbs, H.J.: Engineering properties of expansive clays. ASCE Trans. **121**, 641–677 (1956)

Kumar, A., Walia, B.S., Bajaj, A.: Influence of fly ash, lime and polyester fibers on compaction and strength properties of expansive soil. J. Mater. Civil Eng. **19**(3), 242–248 (2007)

Puppala, A.J.: Fibre and fly ash stabilization methods to treat soft expansive soils. Geotech. Spec. Publ. **112**, 136–145 (2001)

Puppala, A.J., Musenda, C.: Effect of fiber reinforcement on strength and volume change in expansive soils. Trans. Res. Rec. **1736**, 134–140 (2000)

Shanker, N.B., Maruthi, G.: Use of lime-soil piles for in-situ stabilization of black cotton soils. Proc. Indian Geotech. Conf. **1**, 149–153 (1989)

Sharma, R.S.: Mechanical behavior of unsaturated highly expansive clay. Ph.D. thesis, University of Oxford, UK (1998)

Sharma, R.S., Phanikumar, B.R., Rao, B.V.: Engineering behavior of a remolded expansive clay blended with lime, calcium chloride and rice-husk ash. J. Mater. Civil Eng. **20**, 509–515 (2008)

Sivapullaiah, P.V., Subbarao, K.S., Gurumurthy, J.V.: Stabilization of rice husk ash for use as cushion below foundations on expansive soils. Ground Improv. **8**(4), 137–149 (2004)

Viswanatham, B.V.S., Phanikumar, B.R., Mukherjee, R.V.: Effect of polypropylene tape fibre reinforcement on swelling behaviour of an expansive soil. Geosynth. Int. **16**(5), 393–401 (2009)

Feasibility of Utilization of Metalized Plastic Waste in Cohesionless Soil

Siddharth G. Shah[1(✉)], Ankur C. Bhogayata[2], and Sanjay Kumar Shukla[3,4]

[1] Civil Engineering Department, Faculty of PG Studies and Research, Marwadi Education Foundation, Rajkot, Gujarat, India
siddharth.shah@marwadieducation.edu.in
[2] Civil Engineering Department, Faculty of Engineering, Marwadi Education Foundation, Rajkot, Gujarat, India
[3] Civil and Environmental Engineering Discipline, Edith Cowen University, Perth, Australia
[4] VIT University, Vellore, India

Abstract. This paper presents the test results of shear strength of cohesionless soil reinforced with macro-sized metallized plastic waste (MPW). The soil samples were collected from Rajkot city of Gujarat state, India. The objective was to obtain the effects of addition of MPW in cohesionless soil on shear strength variation of soil. MPW was received from a local plastic packaging unit as films and shredded into flakes of 5 mm × 5 mm average size. The flakes were mixed in soil by weight fractions from 0% to 3% with an increment of 0.5%. The direct shear test was performed on each specimen. Soil specimen containing 0% MPW was regarded as reference soil. The test results revealed that inclusion of MPW improved the shear strength of soil for MPW dosage up to 1.5% by weight. Beyond 1.5% of addition of MPW flakes reduced the shear strength. However, the experimental study demonstrated the feasibility for a sustainable utilization of MPW in cohesionless soil. Soil reinforced with MPW could be used for the civil engineering applications where shear strength enhancement is required.

1 Introduction

Solid waste produced by empty food packet creates a big disposal problem in India. Lack of incineration and haphazard disposal in villages and urban area have led to a new problem that cattle's eat it and fall sick. To overcome the disposal of this metallized plastic waste few attempts have been made to utilize the waste in concrete without affecting its mechanical and durability properties. Limited efforts are found in utilizing metallized plastic waste in soil. Reports are available for other wastes like fly ash, foundry waste and rubber tyre wastes incorporated in soil subgrade for highway construction. Few researchers such as Consoli et al. (1998, 2005); Hamidi and Hooresfand (2013) and Park (2009), tried to incorporate the polyurethane fibers along with cement to improve the sandy soil. They conducted UCS and concluded that brittleness of soil is decreased due to the fiber reinforcement in comparison to soil only treated with cement.

S.K. Shukla and E. Guler (eds.), *Advances in Reinforced Soil Structures*,
Sustainable Civil Infrastructures, DOI 10.1007/978-3-319-63570-5_5

The mechanism of how MPW waste will act in soil as fibres or as flakes is a matter of study. In the present study, MPW waste flakes were created and were mixed in soil randomly. The presence of randomly oriented fibres in soil mix influences the macroscopic behaviour of the soil composite which contributes to structural anisotropy. Many researchers have assumed isotropy for simplicity during analysis. In a review of fibre- reinforced soil, Hejazi et al. (2012) reported that soil isotropy was assumed to be sustained by randomly distributed fibre in the soil matrix. However, Diambra et al. (2010) suggested that the preferred plane of orientation of the fiber is sub-horizontal when common methods such as moist tamping or vibration are used in preparing the reinforced sample in the laboratory. Moreover, practical applications often require compaction by rolling of fiber reinforced soil which creates the horizontal bedding plane (Michalowski and Creak 2002). Contribution of fibre to the shear strength is effective when subjected to tension while the fibre plays no role in compression as the fibre may buckle or kink.

In this study, efforts have been made to study the shear response of the soil mixed with metallized plastic waste. This can be an innovative way of achieving a dual benefit of probable advantageous shear strength improvement and mitigating the hazardous effects of metallized plastic wastes, which are the one of the sources for littering and landfill.

2 Materials and Methods

2.1 Metallized Plastic Waste

Metallized plastic used by food packaging industries was obtained from Shri Umiya Plastics – a plastic packaging industrial unit at Shapar-Veraval industrial area near Rajkot city. Metallized plastic film was shredded into flakes/fibres of 5 mm × 5 mm from as shown in Fig. 1. The fibres were mixed in varying fractions from 0% to 3% by weight of soil in increment of 0.5%. Soil containing 0% of MPW was considered as reference soil. General properties of metallized plastic wastes are shown in Table 1.

Fig. 1. Shredded metallized plastic waste (5 mm square, average size)

Table 1. Properties of Metallized Plastic Waste (MPW)

Property	Values	Unit
Resin category	Polythene	–
Plastic type	LDPE	–
Recycling code	4	–
Density range	0.94–1.4	g/cm^3
Thickness	0.08	mm
Water vapor resistance	Good	–
Oxygen permeability	High	–
Elongation	10–30	%
Coefficient of friction	0.45–0.55	–

The MPW are non-degradable and chemically inert material. That's how its disposal problem is severe. If they are added in soil for waste utilization, it could be expected that it will survive for at least six to seven decades without any problem (Narayan 2001); Marsh et al. (2007).

2.2 Cohesionless Soil

The soil considered in this study was collected from river Bhogavo, Gujarat India. The soil particles were of cubic to irregular shape and the average size particle D_{10} was around 1.2 mm. The gradation curve and other engineering properties are given in Fig. 2 and Table 2, respectively.

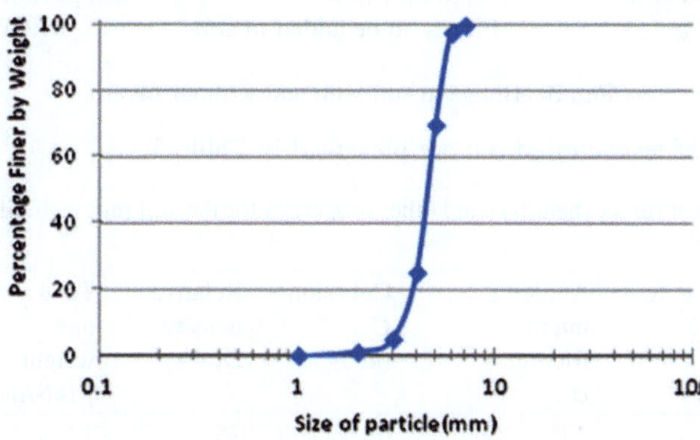

Fig. 2. Particle-size analysis (Gradation curve) of Bhogavo soil considered in this study.

2.3 Sample Preparation

To utilize the plastic waste in sandy soil, the plastic waste was collected from the Rajkot city of Gujarat and it was shredded in to average size of 5 mm × 5 mm in average. Then it was added to the soil by fraction of the weight of the dry sand and mixed thoroughly by hand. The mix prepared was then divided in three equal parts.

Table 2. Engineering properties of Bhogavo soil

Sr. No	Property	Value
1	Angle of internal friction	29^0
2	Cohesion	0 kN/m^2
3	Relative density	75%
4	Total unit weight	17 kN/m^3
5	Moisture content	14%

After that the soil mix was compacted in assembled box by the tamping rod by 25 blows to achieve the desired relative densities. Likewise all the remaining two layers were compacted. After this, the regular tests for shear strength and relative density as per the Indian standards IS 2720 were carried out on the soil. The soil sample and its mixing procedure are given in Fig. 3.

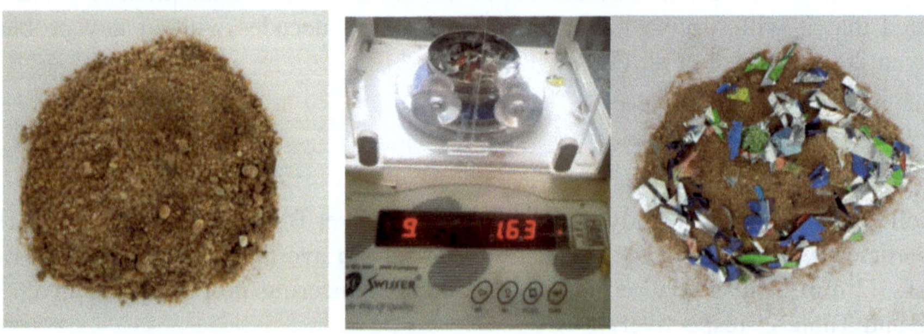

Bhogavo soil | Weighing MPW Fibres to be added in soil | Bhogavo Soil mix with MPW fibres

Fig. 3. Bhogavo soil with and without fibres

The results of tests carried out are presented in Table 3.

Table 3. Results of direct shear test and other properties for the soil mix with different dosage of MPW

Sr. No	Nomenclature to soil mix	Angle of internal friction ϕ	Cohesion C (kPa)	Relative density D_r (%)	Total unit weight γ (kN/m^3)	Moisture content w (%)
1	SWP	29°	0	75	17	14
2	0.5 SMPW	29.3°	0	74	16.9	14
3	1.0 SMPW	30°	0	73	16.8	14
4	1.5 SMPW	30.5°	0	73	16.8	14
5	2.0 SMPW	29.5°	0	72	16.7	14
6	2.5 SMPW	28.8°	0	72	16.7	14
7	3.0 SMPW	28.3°	0	0.71	16.6	14

SWP: Soil without Metalized Plastic waste.
SMPW: Soil with Metalized Plastic waste.

2.4 Methods

In this study, to estimate the effect of addition of metallized plastic waste flakes in sandy soil the shear strength by direct shear test was carried out. As the soil is sandy, so the drainage conditions were not affecting the shear strength and undrained parameters were evaluated. The soil was compacted at relative density of 70% and with its natural moisture content of 14%. The metallized plastic waste fiberes by percentage of the weight of the total soil from 0 to 3% with increment of 0.5% were added in respective tests. The preparation of sample for direct box shear test is shown in Fig. 4.

Soil mix compacted in shear box

Assembled box in the direct shear test apparatus

Fig. 4. Preparation of soil specimen for direct shear box test.

3 Results and Discussion

The soil without metallized plastic waste was considered as reference soil, which had an angle of friction of 29°. When the metallized plastic waste was added to it, the sand particles under the normal load were partially punched in to the flakes, thus the friction between the flakes and sand particles was executed under shear forces. As the normal stress increases, the flakes were under tension to resist the shear forces hence there was an improvement in angle of internal friction of soil. This improvement was not significant but around 5% at 1.5% dosage of MPW addition in soil by weight. As the dosage of the MPW was increased beyond 1.5%, significant volume of MPW was available in the soil mix which reduced the contact surface area between soil particles and plastic surface, and it was observed that two or three flakes were attached to each other and sand particles were not punched in the flakes therefore, the angle of internal friction was reduced and was found in the range of 5% to 7%.

4 Conclusions

In this study feasibility of the addition of the metallized plastic waste in cohesionless soil was attempted. Based on the testing results and analysis, the following general conclusions can be drawn:

1. Addition of metallized plastic up to 1.5% is favourable as it does not affect the engineering properties of sand, instead the shear strength parameter ϕ is increased by 5%.
2. Utilization of MPW in sand is possible as a constituent in reinforced sand. Dosage of MPW at 1.5% by weight of soil will give a considerable volume of MPW, so problem of littering and landfill might be mitigated.

References

Consoli, N.C., Casagrande, M.D., Coop, M.R.: Effect of fiber reinforcement on the isotropic compression behavior of a sand. J. Geotech. Geoenviron. Eng. **131**(11), 1434 (2005). doi:10.1061/ASCE1090-0241

Consoli, N.C., Prietto, P.D., Ulbrich, L.A.: Influence of fiber and cement addition on behavior of sandy soil. J. Geotech. Geoenviron. Eng. **124**(12), 1211 (1998). doi:10.1061/(ASCE)1090-0241

Hamidi, A., Hooresfand, M.: Effect of fiber reinforcement on triaxial shear behavior of cement treated sand. Geotext. Geomembr. (2013). doi:10.1016/j.geotexmem.2012.10.005

Park, S.S.: Effect of fiber reinforcement and distribution on unconfined compressive strength of fiber-reinforced cemented sand. Geotext. Geomembr. (2009). doi:10.1016/j.geotexmem.2008.09.001

Hejazi, S.M., Sheikhzadeh, M., Abtahi, S.M., Zadhoush, A.: A simple review of soil reinforcement by using natural and synthetic fibers. Constr. Build. Mater. (2012). doi:10.1016/j.conbuildmat.2011.11.045

Diambra, A., Ibraim, E., Wood, D.M., Russell, A.R.: Fibre reinforced sands: experiments and modelling. Geotext. Geomembr. (2010). doi:10.1016/j.geotexmem.2009.09.010

Michalowski, R.L., Čermák, J.: Strength anisotropy of fiber-reinforced sand. Comput. Geotechn. (2002). doi:10.1016/S0266-352X(01)00032-5

Narayan, P.: Analysing plastic waste management in India. Case study of Polybags and PET bottles. In: IIIEE Reports, vol. 11 (2001)

Marsh, K., Bugusu, B.: Food packaging—roles, materials, and environmental issues. J. Food Sci. **72**(3), R39–R55 (2007)

Comparison of Geotextile-Reinforced and Geogrid-Reinforced Flexible Pavements by Numerical Analyses

Nadjet Bouacha(✉)

Department of Civil Engineering, Faculty of Science and Technology, University Med Cherif Messaedia Souk-Ahras, Souk-Ahras, Algeria
n.bouacha@yahoo.fr

Abstract. Over the past three decades, geosynthetics have been used successfully around the world in many areas of civil engineering, and are now a well-accepted building material. Their use provides excellent economic alternatives to conventional solutions to many engineering problems. Therefore, students and practicing engineers need exposure to the fundamentals of geosynthetics as a building material. The Geosynthetics is a generic term for all synthetic materials used in conjunction with the soil, rock and/or other-related civil engineering material as an integral part of a project, structure or system. Geomembranes are used to distinguish their sealing qualities or permeability and geotextiles used for their mechanical functions. This study is to analyze the behavior of the pavement structure reinforced with layers of geotextiles. This analysis is done through a numerical modeling with the code PLAXIS V8. The latter is based on the principle of finite elements, this criterion will help us to better understand the behavior of the pavement structure and the ground vis-à-vis the parameter analysis of stress and strain. The principle of this analysis is based on a comparison designed pavement with and without geotextiles and will focus on the radial stresses settings, vertical stresses and displacements for two types of materials processed bitumen treated materials bitumen structures and materials treated with hydraulic binders.

Keywords: Flexible pavement · Hydraulic binders · Stress · Displacement · Geotextile · Plaxis · Modelling

1 Introduction

The pavement body is a multilayer structure. Its overall behavior depends on the nature of the materials that compose it, and their importance to achieve a good bond at the interface between pavement layers throughout his life. The stress caused by the traffic and environmental conditions are the main causes of damage to the pavement layer, leading to more degradation modes, night security, and quality of service, reduce maintenance costs a building pavement is needed and on the other hand speak of geotextiles as strengthening of the pavement structure solution and begin a numerical modeling with a code based on the principle of finite elements. The latter is based on the principle of finite elements, this criterion will help us to better understand the

S.K. Shukla and E. Guler (eds.), *Advances in Reinforced Soil Structures*, Sustainable Civil Infrastructures, DOI 10.1007/978-3-319-63570-5_6

behavior of the body floor and vis-à-vis ground analysis of the parameters of stress and strain (Jeuffroy 1911). This study will help us to analyze and understand the behavior of the body influence pavement geotextile towards the vertical stress parameters; radial strain and displacement (January and Mamadou 2007; Boussinesq 1885; Peyronne and Caroff 1984).

A geosynthetic is the generic term for a product of which at least one of constituents is based on synthetic or natural polymer, in the form of band, or three-dimensional structure, used in contact with the ground or with other materials in the fields of geotechnics or civil engineering. Geosynthetics are classified into two main families:

1. Permeable products: geotextiles and geotextile-related products,
2. Essentially impermeable products: geomembranes and related products geomembranes (Bhandari and Han 2010; Alexiew et al. 2010).

The type of Geosynthetics utilised in this work is the Geotextile, they are products from the textile industry, from natural origins (fibers of cotton and jute) or synthetic (polyester, polyethylene, polypropylene, rarely polyamide). Products related to geotextiles are mainly geogrids, Geobags, geotubes, polymer geocontenters. Geotextiles are used and better known as geo membranes used in particular for waterproofing works. In all the works, geotextiles meet at least five basic functions: the separation, filtration, drainage, reinforcement and the fight against erosion. Geotextiles are classified according to their structure, that is to say, depending on the manufacturing process which, from polymeric fibers (mainly polypropylene) yielded a finished material. These "families" have names from the textile industry. Thus, the geotextilles can be woven geotextille products from son monofilaments, multifilaments son, or tape; nonwoven geotextiles can be needled or thermally bonded, or even knited.

2 Approach Adopted

In this work we tried to follow the successive steps to reach the objectives which are:

1. The influence of the behavior of geotextile flexible pavements.
2. The influence of geotextile on two types of body flexible pavements: *Structure Treated with Bitumen (STB) and* Structure with materials hydraulic binders (SMHB). For the PLAXIS software to do the calculations correctly and completely, we must take it all the project data.

2.1 Assumptions

The assumptions for modeling two structures are summarized in the following points (Quang and Tran 2004):

- The deformations are considered flat.
- All interfaces are glued except base courses are half-pasted.
- Material properties of each layer are homogeneous,

- The layers are infinite in lateral directions
- Solutions constraints are characterized by two main properties of each layer "of the fish coefficient and the elastic modulus E".

2.2 Geometry

The structures are modeled by planar geometric patterns in two dimensions (2D) 35 m wide and 15 m deep. An example of the models is shown in the following Fig. 1.

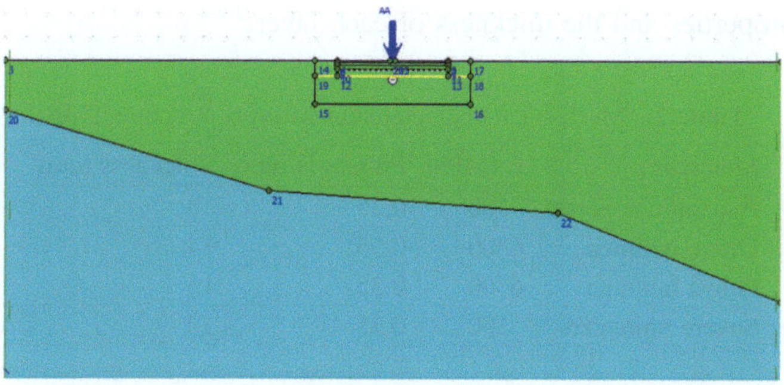

Fig. 1. Physical model

2.3 Properties of Soil Layers

The massif is composed of two types, characteristics are shown in Table 1.

Table 1. Properties of soil layers integrated in the model

Parameters	Name	Soil 1	Soil 2		Unit
		Mohr coulomb			
Type of behavior		Drained	Drained	Drained	
Dry unit weight	$\gamma unsat$	18	13	17	kN/m^3
Wet volume weight	γsat	21	17	19	KN/m
Poisson's ratio	*V*	0	0	0	-
Horizontal permeability	Kx	0	0	0	m/day
Vertical permeability	Ky	50	50	60	m/day
Young's modulus	Eref	0.35	0.35	0.35	Mpa
Cohesion	c_{ref}	30	5	5	-
Friction angle	Φ	25	20	35	kN/m^3
Angle of expansion	ψ	0	0	0	°
Rigidity factor interface	Rinter	Rigid	Rigid	Rigid	°

2.4 Pavement Dimensions

For the materials treated in the bitumen structure:

- Width: 5 m
- Thickness: 0.67 m

The material structure hydraulically bound:

- Width: 5 m
- Thickness: 0.82 m

Modelled roads are broken into four layers, Tables 2 and 3 summarize the mechanical properties and the thickness of each layer.

Table 2. Mechanical characteristics of layer structure TMB

Materials	E (MPa)	Poisson's ratio	Thickness (cm)
Asphaltic concrete	3600	0.35	8
Grave bitumen	6300	0.35	14
Grave bitumen	6300	0.35	15
Severe untreated	500	0.35	30

Table 3. Mechanical properties of the structural layers HTMB

Materials	E (MPa)	Poisson's ratio	Thickness (cm)
Asphaltic concrete	3600	0.35	10
Dairy grave	23000	0.25	21
Dairy grave	23000	0.25	21
Severe untreated	500	0.35	30

¬ Properties of geotextiles:
Geotextiles used are woven geotextiles with the properties listed in Table 4:
¬ Loading:

Table 4. Characteristics of geotextile

Parameter	Name	Value	Unit
Axial stiffness	EA	$6.87 * 10^{05}$	Kn/m
Behavior		Elastique	Mm
Thickness under 2 kPa		1	g/m^2
Area weight		200	kn/m^2
Tensile strength		16	Mm
Dynamic perforation (cone drop)		17	Kn
Static punching		0.9	Kn
Permeability		0.045	m/s
Filtration opening		75	Mm

The reference axle single wheel isolated 130 km. In the design, there is a reference half-axle and the modeled as a uniformly distributed load of 0.662 MPa to a disc of 0.125 m radius (Peyronne and Caroff 1984).

3 Data to Introduce

3.1 Materials Treated with Bitumen (TMB)

3.1.1 Properties of Materials Used for the Body Pavement in Tables 5 and 6

Table 5. Characteristic material used in the body of shoes

Parameters	Name	AC	GB	GB	Unit
		Linear elastic			
Type of behavior		Non porous	Non porous	Drained	
Dry unit weight	$\gamma unsat$	22	22	22	kN/m^3
Wet volume weight	γsat	-	-	22.8	KN/m^3
Horizontal permeability	Kx	-	-	-	m/day
Vertical permeability	Ky	-	-	-	m/day
Young's modulus	Eref	4000	7000	500	Mpa
Poisson's ratio	V	0.35	0.35	0.35	-
Cohesion	c_{ref}	-	-	-	kN/m^3
Friction angle	Φ	-	-	-	°
Angle of expansion	ψ	-	-	-	°
Rigidity factor interface	Rinter	Rigid	0.8	Rigid	

Table 6. Characteristics of materials used in the body of shoes

Paramètres	Nom	BB	GL	GU	Unit
Type Model	*Model*	élastique linéaire			
Type of behavior	*Type*	Non	Drained	Drained	
Dry unit weight		poreux	23	22	
Wet volume Weight		22	23.8	22.8	
horizontal	$\gamma unsat$	-	-	-	kN/ m^3
permeability	γsat	-	-	-	kN/ m^3
vertical permeability	*Kx*	-	23000	500	m/day
Young's modulus	*Ky*	4000	0.25	0.35	m/day
Poisson's ratio	*Eref*	0.35	-	-	Mpa
Cohesion	v	-	-	-	-
Friction angle	c_{ref}	-	-	-	kN/m3
Angle of expansion	φ	-	0.8	Rigid	°
Rigidity factor	ψ	Rigid			
interface	*Kinter*				

3.1.2 Mesh

The generation of the structure with materials treated with bitumen MTB model mesh is made by 15-node elements. The number of elements is 261 elements, and the number of nodes is 2209 nodes. A possibility of mesh refinement obtained may be carried out with the PLAXIS software, where the final number of elements is 551 and the number of nodes is 4581 nodes. Figure 2 shows the mesh mad (PLAXIS V8).

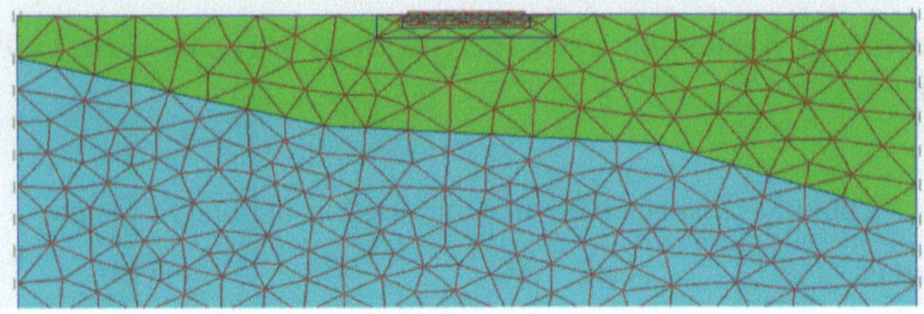

Fig. 2. Final mesh structure

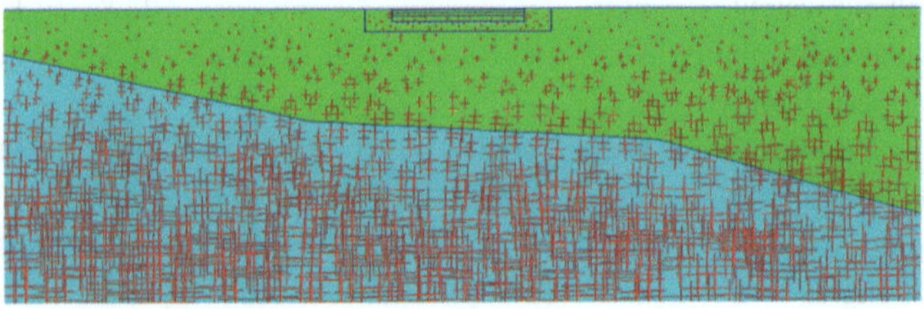

Fig. 3. Initiation effective stress

3.1.3 Initial Conditions

The initial conditions require the generation of initial stresses (Fig. 3):

For the calculation of the initial stresses, disable and structural elements the pavement element created by default.

Is generated by taking the initial constraint values Ko automatically proposed according to the formula Jaky. We keep the weight of the soil to 1, which corresponds to a total application of gravity (Plaxis, V8).

3.1.4 Calculation Procedures

- The Calculation of the reference model is defined in 3 stages in the order as follows:

  Phase 0 (initial phase):

- Initiation of constraints (K_0 procedure); the initial effective stress is determined.

  Phase 1:

- Establish a relaxing pavement layers directly on the sub grade.
- Activation of the charge of a single tire with a pressure value is = 0.662 MPa.

Phase 2:

- Take phase 0 as a starting phase
- To restore the structure by placing the pavement layers, this time on a thick layer of fill of 1 m
- Activation of the load of the single tire with a value of 0.662 Mpa.

Phase 3:

- Take phase 0 as a starting phase
- Add a layer of geotextile under the layer of GNT
- Activate the load the tire.

3.2 Structure Treated Materials Hydraulic Binders (SMHB)

3.2.1 Ownership of the Materials Used for the Body Pavement

The following table shows the characteristics of the materials used in the body of shoes.

3.2.2 Mesh

The generation of the MTB model mesh is made by 15-node elements. The number of elements is 259 elements, and the number of nodes is 2193 nodes. A possibility of mesh refinement obtained may be carried out with the PLAXIS software, where the final number of elements is 549 and the number of nodes is 4565 Nodes presented in Fig. 4.

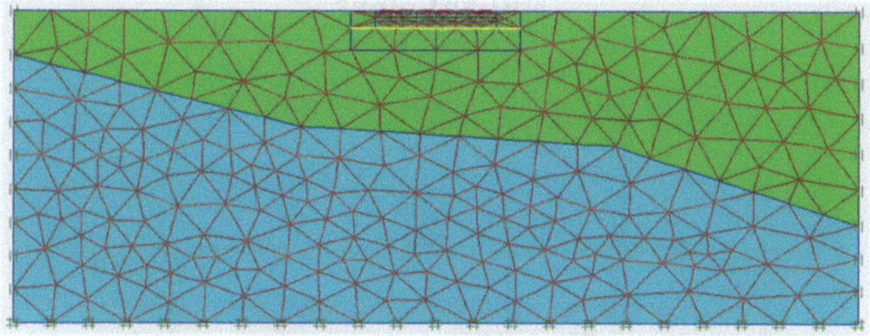

Fig. 4. Mesh generation

3.2.3 Initial Conditions

Procedure same as previously described structure (Figs. 5 and 6).

3.2.4 Calculation Procedures

- The Calculation of the reference model is defined in 3 stages in the order as follows:

  Phase 0 (initial phase):

- Initiation of constraints (K0 procedure); the initial effective stress is determined.

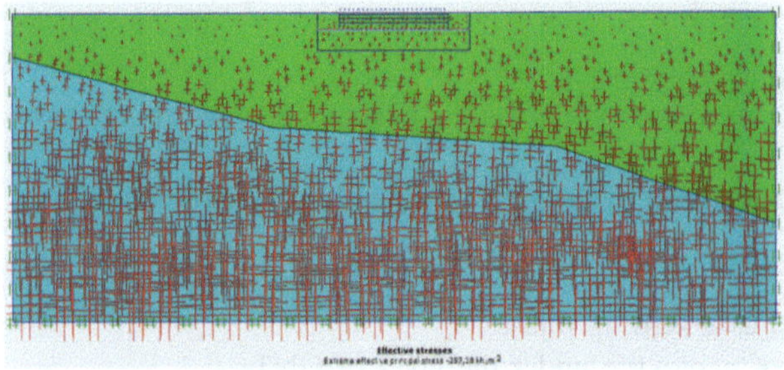

Fig. 5. Initial stress

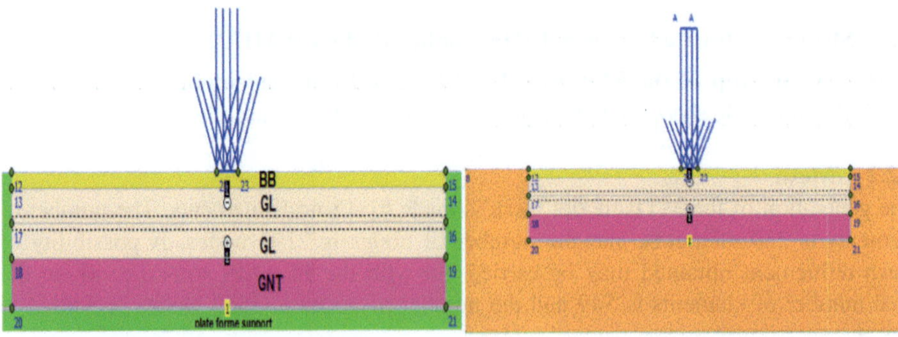

a. Initial STR1structur **b. Structure of STR2**
embankment

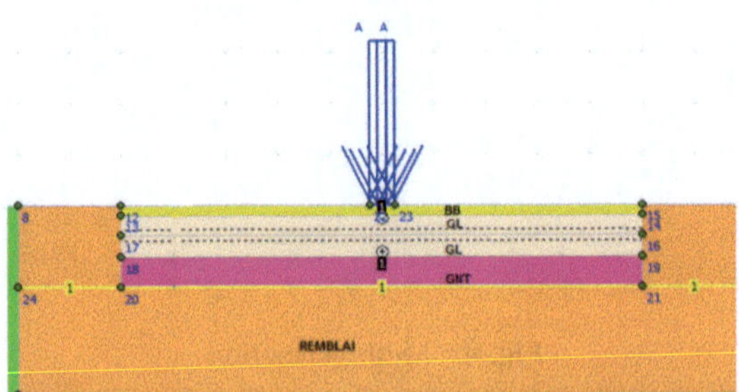

c. Structure reinforced with geotextile STR3

Fig. 6. Types of structures

Phase 1:

- Establishment of pavement layers resting directly on the subgrad.
- Activation of the charge of a single tire with a pressure value is = 0.662 Mpa.

Phase 2:

- Take phase 0 as a starting phase
- To renovate the structure by placing the pavement layers this time on a thick layer of fill 1 m
- Activation The charge of a single tire with a value of N = 0.662 Mpa.

Phase 3:

- Take phase 0 as a starting phase
- Add a layer of geotextile under the layer of GNT
- Activate the load the tire.

4 Results and Analysis

Modeling has gone through various stages, and according to the results found guidance was performed to achieve the best results. These key steps are:

- Pavement located on natural ground. "STR1" Pavement located on natural ground strengthened by an embankment "STR2".
- Consolidated pavement located on natural ground by an embankment and reinforced by geotextile "STR3".
  - The comparison is based on the three parameters for both TMB and STHB structures.
  - to.

    a. the vertical displacement
    b. the radial stress;
    c. the vertical constraints;

4.1 TMB Model

(1) Displacement

Figure 7 shows the behavior of the constituent layers of the pavement and the ground under the effect of a charge of a single tire.

As regards the displacement can be estimated three intervals which are: On the surface of the three curves start with a maximum displacement worth 6.19 to 1 STR (initial structure); 5.48 to STR2 (structure embankment); and 4.87 for STR3 (structure reinforced by géotextile). The displacement will decreases approaching the géotextile web; below the water table and the two curves STR2 STR3 follow the same pace; time that STR1 curve follows a path aggressive; Until the three curves meet at the same point. Beyond this point the three curves follow the same pace; and displacement of values weakens until they cancel out.

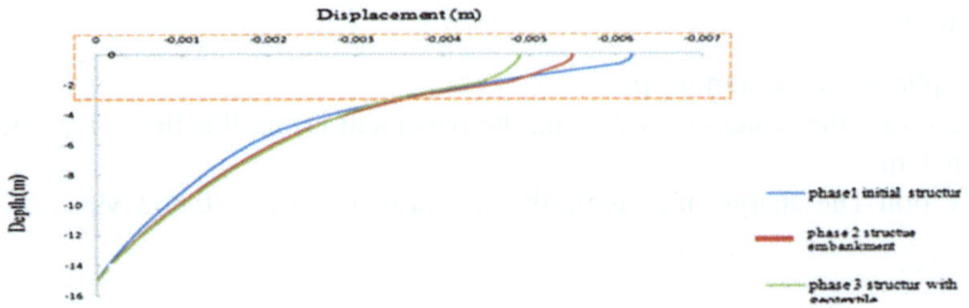

Fig. 7. Vertical displacement for the three structures MTB

(2) Radial Stress

Three structures STR1, STR2, STR3. From Fig. 8 we see a good correlation between the three curves (STR1, STR2, STR3) with remarkable gap with STR3 curve, this difference shows a good distribution of the radial stress distribution at the surface the load. The vertical stress note for the three profiles digressive distribution of vertical stress; the load is maximum at the surface.

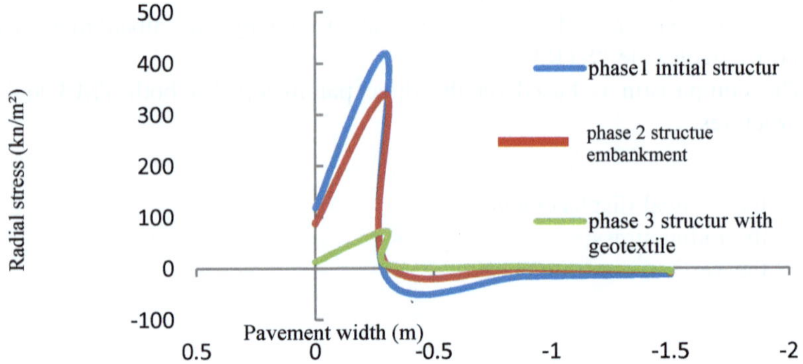

Fig. 8. Comparison of the distribution of the radial stress for the three structures STR1, STR2, STR3.

The load is picked up by the top layer of the floor and then diffuses into the other layers of floor to the distribution in the soil until the cancellation. The geotextile is influenced by the vertical stress at the base layer.

4.2 Structure Bills of Materials for Hydraulic Binders (STHB)

(1) Vertical Displacement

From Fig. 9, we see that the three curves follow similar appearance to those of bitumen processed structure, but with a minimum displacement and close values (4.87 mm for STR1, STR2 4.52 mm, 4.32 mm STR3).

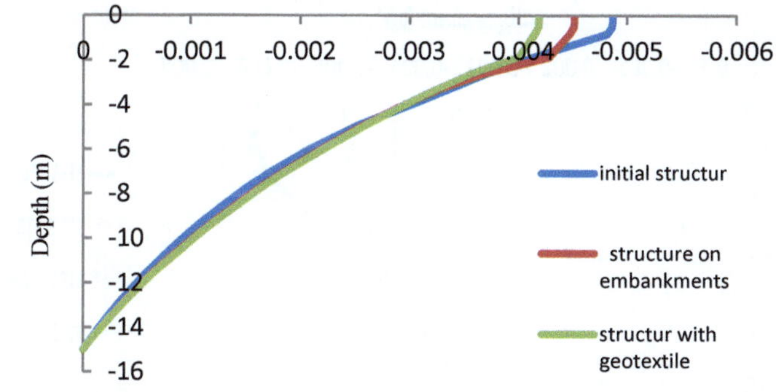

Fig. 9. Comparison of traveling between the three structures STR1, STR2, STR3 hydraulically bound

The radial stress in the structures treated with hydraulic binders is manifested by a significant compression that develops at the surface layer. Hose compressive stresses develop tensile stresses between the interfaces of the upper layers of the pavement. We notice the same behavior for the three structures, with minimum values for the pavement structure with geotextile. Through Fig. 10 we see that the three curves are perfectly superimposed on the body of the pavement. A slight difference occurs at ground level (Figs. 11, 12 and 13).

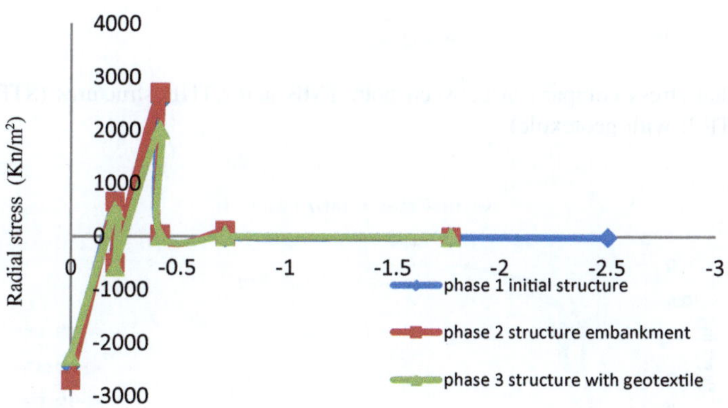

Fig. 10. Comparison of radial stress for the three structures STR1, STR2, STR3 treated with hydraulic binders.

4.3 Comparison Between Both Structure Treated with Hydraulic Binders (STHB) and Structure Treated Bitumen (STB) Models

(1) Vertical Displacement

We note that the use of geotextile greatly reduces the displacement for the two types of structures. Also we note that travel for STHB structures are smaller than for MTB structures.

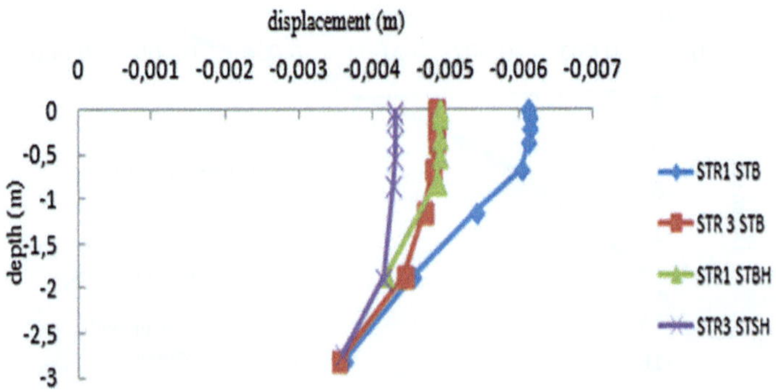

Fig. 11. Shows the comparison of displacement profiles for both types of STB and STHB structures with and without webs of geotextile.

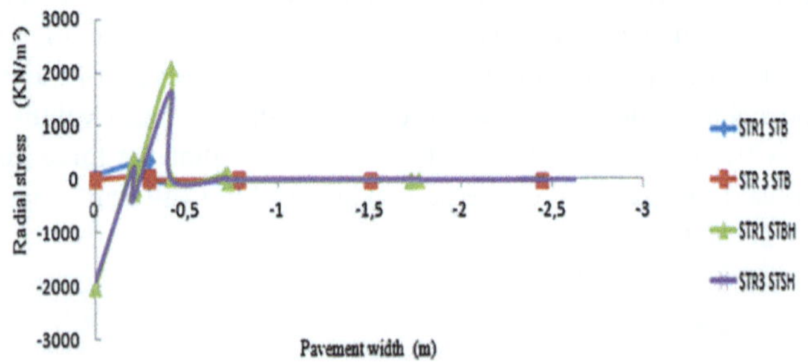

Fig. 12. Radial stress comparison between both TMB and STHB structures (STR 1: without geotextile, STR3: with geotextile)

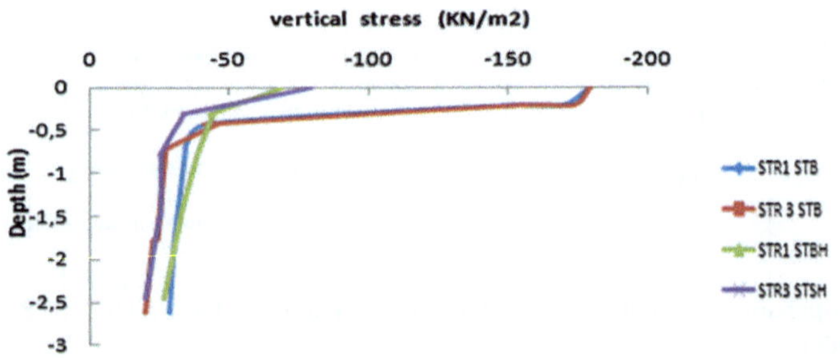

Fig. 13. Vertical stress comparison between both STB and STHB structures (STR 1: without geotextile STR3: with geotextile)

(2) Radial Stress

The comparison shows a very big difference between the results of the MTB model and those of MTLH model; wave propagation in radial stresses, deep below the body of the floor, almost the same pace for all the observed curves. therefore the interpretation is made on the [0 −2.68] which is representative for the reinforcement with a geotextile mat on this section we can see that the difference is more marked in the coating zone, or coercion takes very deferential values: Starts with negative values for MTLH and positive values for the MTB in two different paces.

(2) Vertical Stress

The vertical stress is maximum at the surface (the contact pressure). In depth, this constraint decreases almost linearly through the coating, the structure undergoes STHB larger vertical stress values than those suffered by the STB and takes a STHB the appearance of a small slope to the level or we placed the geotextile from the coating-geotextile interface fourths curves follow the same path from the ground pavement in the body, or the vertical stress continues to decrease and tends to be canceled on the basis of the model.

5 Analysis

By paying attention to the location area of the geotextile, we note that the curves have experienced remarkable changes. However the two structures and STB sudden STHB different results:

1. Structure treated with Bitumen: concerning this structure we have:
    - A minimum shift large reduction in radial stress
    - Not a large difference of the vertical stress before and after the addition of the geotextile.
2. And for Structure Treated with Hdraulic Bituminen was:
    - Small decrease of displacement.
    - The addition of the geotextile seems has no influence to the decrease in radial and vertical constraints.

Ultimately, we can attribute the great difference between the two structures to the quality of materials used and the proportionality of the geotextile with each structure, and believe that with the use of a geotextile STB structure the same can be achieved STHB performance of a structure subjected to the same conditions (traffic, climate, soil bearing). In the end we say that the analysis performed is only approximate because many input parameters were approached (cohesion, friction angle, etc.).

6 Conclusions

The principle of building roadways as for other civil engineering structures is to determine the stresses caused by a vehicle and compare them with the parameters limit values of the various constituent materials of the structure. This level of stress is evaluated by a mechanical model of the pavement. The latter that researchers are trying to develop it to make it more representative of physical reality.

Especially as the theory assumes many simplifying assumptions. The development of the mechanical model was not possible without the development of the numerical model and the widespread use of digital computers which helped to solve very complicated problems.

Modelling in the field of road is mediocre, given the complexity of the structure and the many parameters involved "traffic, climate, soil, materials, etc."

The objective of this work was to create a model that takes into account body composition of the roadway, the insertion of the geotextile in two types of structure, treated with bitumen and the other treated with u hydraulic binder, with a recent numerical tool scientifically. Our choice is fixed on the PLAXIS V8 software. The latter is based on the principle of finite elements; this test will help us to better understand the behavior of the body of the pavement and analysis vis-à-vis the stress and strain parameters behaviour (Alexiew and Hangen 2013).

Through this study, we can conclude is that:

The location of géotextille tablecloths in the flexible pavement structure influences in a remarkable way of moving Rating Decrease. Similarly, the radial stress influence is remarkable.

The geotextile is influenced by the vertical stress at the core layer. The use of geotextile can significantly reduce the displacement for the two types of structures (Arab 2015).

Travel for SMHB structures are weaker than for STB structures. The comparison shows a very big difference between the results of the model and those of TMB STHB model; the spread of radial stress waves are different for the different behavior of the materials treated with binders and materials treated with hydraulic bitumen.

The strengthening effect in the traffic lanes structures are generally, but only under certain conditions of deformation and interface (Pameira 2009). The use of geosynthetics can effectively improve the deformation behavior of road structures. In the road sector, studies on the two-dimensional reinforcement sheet (geotextiles, geogrid, geocomposite and a single table or multi-table) do not yield easily generalizable. However, the results are very satisfactory in terms of extending the service life, reduce the appearance.

References

Bhandari, A., Han, J.: Investigation of geotextiles-soil interaction under a cyclic vertical load using the discrete element method. Geotext. Geomembr. **28**(1), 33–43 (2010)

January, B., Mamadou, G.: Rehabilitation and maintenance of flexible pavements by hot recycling of bituminous materials in situ. Theme ingenierat, Cheikh Anta Diop University in Dakar-Senegal (2007)

Boussinesq, J.: Application of potential in the study of equilibrium and motion of elastic body, Paris, Gauthier Villars (1885)

Peyronne, C., Caroff, G.: Road course. National School of Bridges and Floor Presses (1984)

Alexiew, D., Hangen, H.: Design and construction of high bermless geogrid walls in a problematic mountainous seismic region in Bulgeria. In: Proceedings of the 18th Congress on International Soil Mechanics and Geotechnical Engineering, Paris (2013)

Alexiew, D., et al.: Twenty eight meters high geogrid reinforced embankments as flexible solutions in problematic hillsides: project Trieben-Sunk, Austria. In: 9th International Conference on Geosynthetics, Brazil (2010)

Pameira, E.M.: Soil geosynthetic interaction: modelling and analysis. Geotext. Geomembr. **27**(5), 368–390 (2009)

Jeuffroy, G.: Pavement Design, 2nd edn. National School of Bridges and Roads (1911). Guidance on design and construction of earthworks, France, March 2007

Mehrez Khemakhem, M.R.: Testing of geotechnical road: a comparative study between the standards, ISET Sfax-ATMS-AIT

Quang, Q., Tran, D.: Model simplifies for multilayer cracked pavements. Ph.D. thesis, National School of Bridges and Roads (2004)

Arab, R., Durand, R., Benhouhou, M., Mamouni, R., Foura, M.: Glissements de terrains et apport des géosynthtiques, Colloque – ASAG, 28–29 janvier 2015, Alger Algerie (2015)

Thanh, T.: Exploiting marine and river sediments in road building. Ph.D. thesis, University of Artois-dauoi (2009)

Study on Square Footing Resting on Prestressed Geotextile Reinforced Sand

S. Kumar[1(✉)], C.H. Solanki[1], J.B. Patel[1], P.B. Sudevan[1], and P.M. Chaudhary[2]

[1] Applied Mechanics Department, S. V. National Institute of Technology Surat, Surat, Gujarat, India
{skumar, chs, pjb}@amd.svnit.ac.in, priyabeenasudevan@gmail.com
[2] R & B Division, Government of Gujarat, Nanpura, Surat, India
se_rnb_srt@yahoo.com

Abstract. This paper presents the results of laboratory model tests carried out on square footing resting on geotextile reinforced sand. The model steel tank of size 120 cm × 50 cm × 50 cm and square footing of size 10 cm are used. The effect of reinforcement with geotextile of sizes 20 cm × 20 cm, 30 cm × 30 cm and 40 cm × 40 cm below footing at different depth of placement were studies through a series of laboratory model tests. The effects of prestressing the geotextile on bearing capacity improvement and settlement reduction of a reinforced sand bed are also being investigated. The study also highlights the effect of size of geotextile and placement of geotextile below footing on load-settlement characteristics.

Keywords: Bearing capacity ratio · Settlement reduction ratio · Square footing · Geotextile · Prestressing

1 Introduction

Geosynthetics soil reinforcement such as geotextiles, geogrides and geocomposites have beneficial effects on bearing capacity and settlement of shallow foundations. Considerable experimental research has been reported to study the behaviour of footing resting on geosynthetic reinforced bed (Adams and Colin 1997; Basudhar et al. 2007; Boushehrian and Hataf 2003; Guido et al. 1985; Khing et al. 1993; Lackner et al. 2013; Latha and Somwanshi 2009; Lovisa et al. 2010; Sitharam and Sireesh 2004; Tafreshi and Dawson 2012; Yasrobi et al. 2009). From the studies reported in the literature it has been observed that there is a substantial increase in bearing capacity of foundation reinforced with geosynthetics and settlement of foundation also decreases. For the maximum improvement in bearing capacity, different researchers have given different view about following design parameter. (a) u = depth of first layer of reinforcement below footing base. Value of u/B varies from 0.175 to 0.5 (B = width or Dia. of footing) (b) z = vertical spacing between reinforcement layer. Value of z/B varies from 0.2 to 0.46 (c) b = width of reinforcement layer. Value of b/B varies from 2.5 to 4.0. (d) N = No. of reinforcement layers. Value of N varies from 3 to 5. Lovisa et al. (2010)

S.K. Shukla and E. Guler (eds.), *Advances in Reinforced Soil Structures*, Sustainable Civil Infrastructures, DOI 10.1007/978-3-319-63570-5_7

conducted laboratory model tests and finite element analyses on a circular footing resting on sand reinforced with geotextile to study the effect of prestressing the reinforcement. The prestressing force applied was equal to 2% of the allowable tensile strength of the geotextile. They observed that the addition of prestress to reinforcement resulted in significant improvement in the load bearing capacity and reduction in settlement of foundation. Lackner et al. (2013) conducted about 60 path controlled static load displacement tests and 80 cyclic load displacement tests to determine the load-displacement behavior of prestressed reinforced soil structures. They concluded that prestressing the reinforcement improves the load displacement behaviour of reinforced soil structures. Also rather than a circular footing, square or rectangular footings are commonly used. Hence in this investigation an attempt is made to evaluate the effects of prestressing of reinforcement in improving the bearing capacity of square footings supported on geotextile reinforced granular beds.

2 Experimental Investigation

The experimental program reported herein, that involves small scale model test was carried out using a test facility in the Geotechnical Laboratory of Applied Mechanics Department at SVNIT Surat, India. Details of the experimental test program, material used, test procedure and analysis of the test results of model studies on load bearing capacity and settlement behavior of square footings resting on a geotextile-reinforced sand bed and prestressed geotextile-reinforced sand bed are presented below.

2.1 Materials

The material used for granular bed is fine sand and is locally available soil known as Panna sand. The grain size distribution curve of sand is shown in Fig. 1 and properties of sand are given in Table 1. The reinforcement used is geotextile and its properties are given in Table 2.

2.2 Test Setup

The model test was performed in a steel tank of dimension 1200 mm length × 500 mm width × 500 mm depth. The model footing is a mild steel plate of size 100 mm × 100 mm and 25 mm thickness. The square footing was loaded by hand operated gear arrangement system supported against a reaction frame. The load is measured with the help of Load cell and deformation by using two LVDT (Linear variable differential transformer) placed opposite to each other as shown in Fig. 3. The schematic view of test apparatus is shown in Fig. 2. The photograph of experimental set up is shown in Fig. 3.

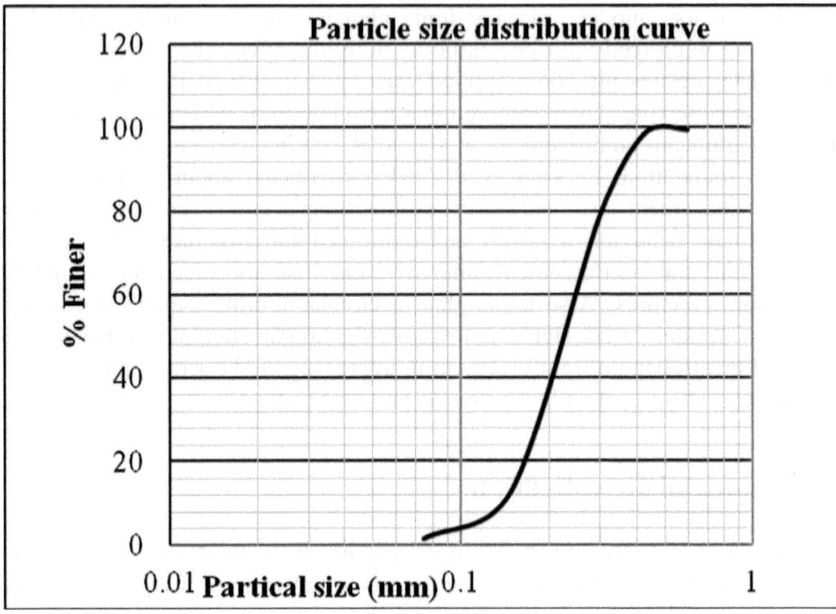

Fig. 1. Grain size distribution curve for the sand

Table 1. Properties of sand

Property	Value
Specific gravity	2.60
Maximum dry unit weight (kN/m^3)	17.3
Minimum dry unit weight (kN/m^3)	14.2
Dry unit weight during test (kN/m^3)	15.5
Effective grain size, D_{10} (mm)	0.14
D_{30} (mm)	0.19
D_{60} (mm)	0.25
Coefficient of uniformity (C_u)	1.78
Coefficient of curvature (C_c)	1.03
Friction angle (ϕ°)	30°
Cohesion, C (kPa)	0

2.3 Preparation of Test Bed

The sand bed is prepared in tank using sand raining technique to achieve required density of sand in each layer. The sand bed is prepared in layers of 50 mm. The sand is filled up to bottom layer of reinforcement. The reinforcement is then placed with its centre exactly below the footing. The geotextile of different size (2B × 2B, 3B × 3B and 4B × 4B) are placed at a depth of 0.1B, 0.2B, 0.3B, 0.4B and 0.5B below footing and then prestress load (2% of the allowable tensile strength of the geotextile) is

Table 2. Properties of geotextile

Property	Value
Mass per unit area, (g/m^2)	147
Thickness, (mm)	1.35
Tensile strength, MD (kN/m)	30
Tensile strength, CD (kN/m)	29
Tearing strength, MD (N)	612
Tearing strength, CD (N)	475
Puncture strength, (N)	637
Burst strength, (N)	290

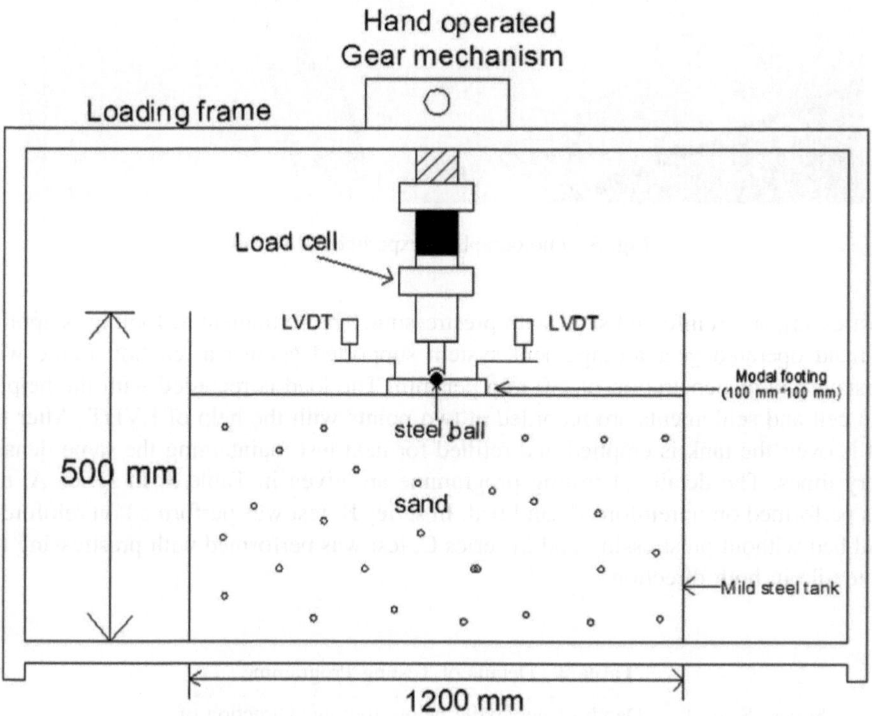

Fig. 2. The schematic view of test apparatus

applied in both direction and is distributed over three pulleys. Then sand above the reinforcement is placed up to footing level and then model footing, Load cell and LVDT are placed.

2.4 Testing Procedure

After the preparation of sand bed, the footing is placed exactly at the centre of geotextile. The tests have been performed for unreinforced sand, reinforced sand without

Fig. 3. Photograph of experimental set up

prestressing and reinforced sand with prestressing. The settlement to footing is applied by hand operated gear arrangement system supported against a reaction frame with constant rate of penetration of 1.0 mm per min. The load is recorded with the help of load cell and settlements are recorded at two points with the help of LVDT. After the test is over, the tank is emptied and refilled for next test maintaining the same density every times. The details of testing programme are given in Table 3. In series A, test was performed on unreinforced sand bed. In series B, test was performed on reinforced sand bed without prestressing and in series C, test was performed with prestressing the geotextile in both direction.

Table 3. Details of Testing Programme

Series	Size of geotextile	Depth of geotextile below footing	Direction of prestress
A	-	-	-
B	2B × 2B	0.1B, 0.2B, 0.3B, 0.4B & 0.5B	-
	3B × 3B	0.1B, 0.2B, 0.3B, 0.4B & 0.5B	-
	4B × 4B	0.1B, 0.2B, 0.3B, 0.4B & 0.5B	-
C	2B × 2B	0.1B, 0.2B, 0.3B, 0.4B & 0.5B	Biaxial
	3B × 3B	0.1B, 0.2B, 0.3B, 0.4B & 0.5B	Biaxial
	4B × 4B	0.1B, 0.2B, 0.3B, 0.4B & 0.5B	Biaxial

3 Results and Discussions

Pressure v/s normalized settlement (S/B %) curve for geotextile of size 2B × 2B placed below footing at the depth of 0.1B to 0.5B, without prestressed and with prestressed are shown in Figs. 4 and 5 respectively. From Figs. 4 and 5, it may be concluded that for geotextile of size 2B × 2B, the optimum depth of placement for maximum improvement in bearing capacity is 0.2B below footing for both the cases.

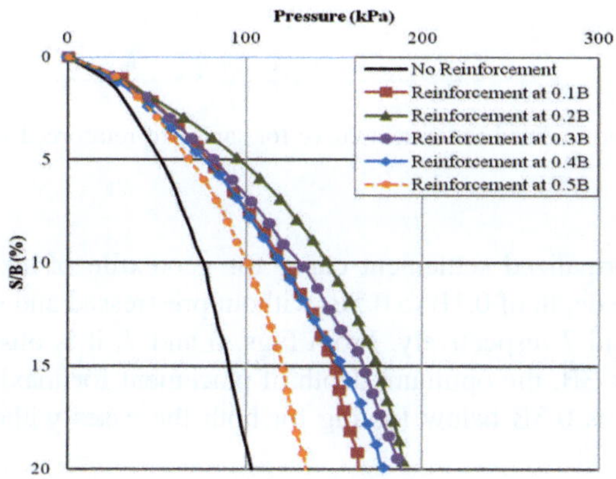

Fig. 4. Pressure v/s normalized settlement curve for sand bed reinforced with geotextile of size 2Bx2B

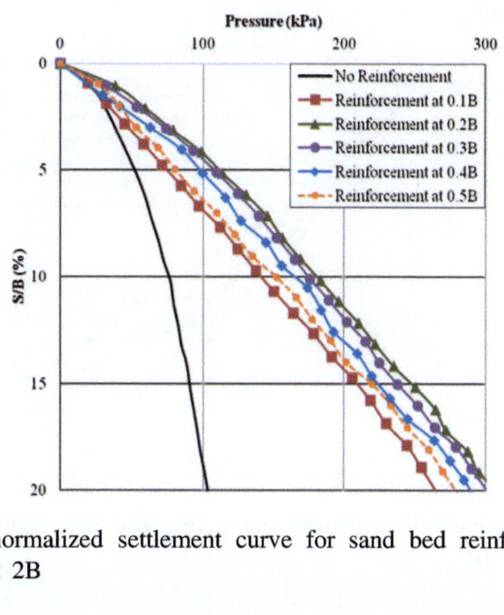

Fig. 5. Pressure v/s normalized settlement curve for sand bed reinforced with prestressed geotextile of size 2B × 2B

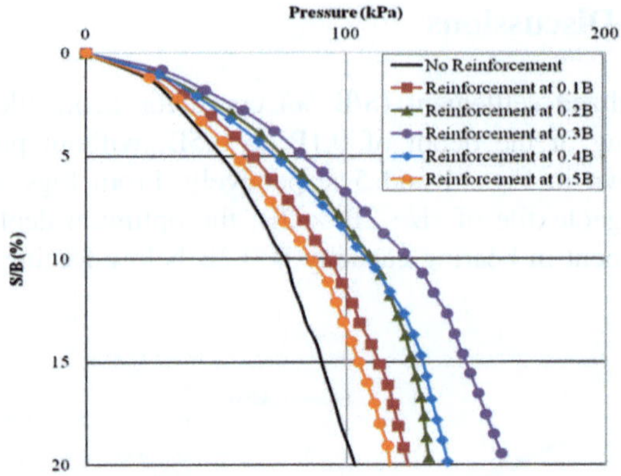

Fig. 6. Pressure v/s normalized settlement curve for sand bed reinforced with geotextile of size 3B × 3B

Pressure v/s normalized settlement curve for geotextile of size 3B × 3B placed below footing at the depth of 0.1B to 0.5B, without prestressed and with prestressed are shown in Figs. 6 and 7 respectively. From Figs. 6 and 7, it is observed that for geotextile of size 3B × 3B, the optimum depth of placement for maximum improvement in bearing capacity is 0.3B below footing for both the cases without prestressed and with prestressed.

Pressure v/s normalized settlement curve for geotextile of size 4B × 4B placed below footing at the depth of 0.1B to 0.5B, without prestressed and with prestressed are shown in Figs. 8 and 9 respectively. From Figs. 8 and 9, it may be concluded that for

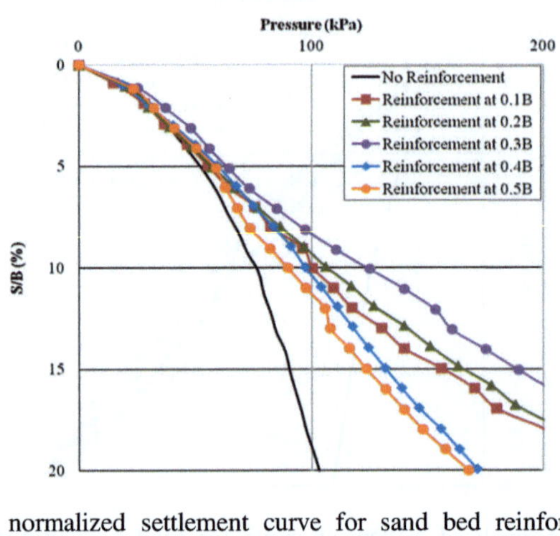

Fig. 7. Pressure v/s normalized settlement curve for sand bed reinforced with prestressed geotextile of size 3B × 3B

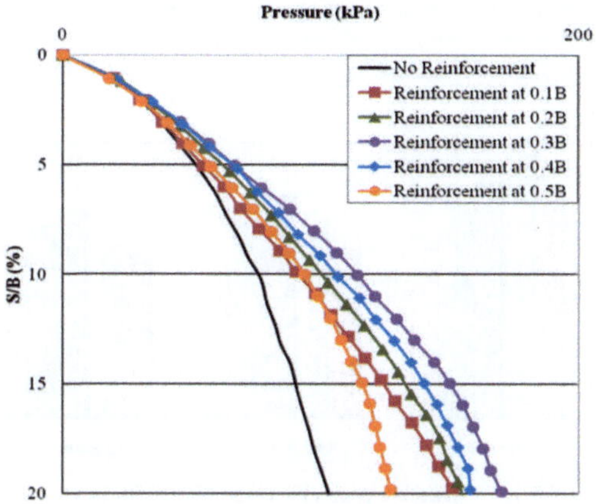

Fig. 8. Pressure v/s normalized settlement curve for sand bed reinforced with geotextile of size 4B × 4B

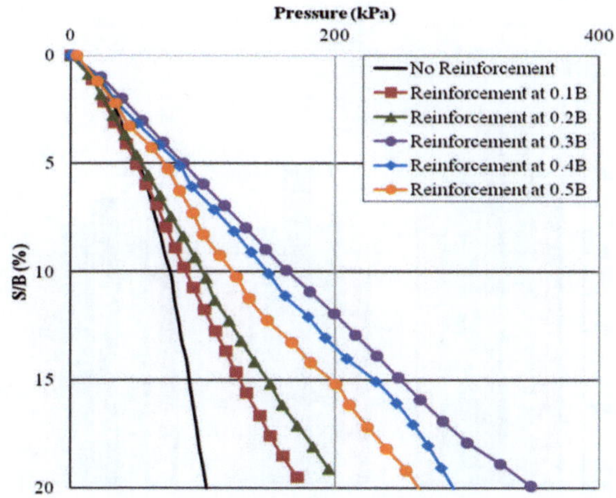

Fig. 9. Pressure v/s normalized settlement curve for sand bed reinforced with prestressed geotextile of size 4B × 4B

geotextile of size 4B × 4B, the optimum depth of placement for maximum improvement in bearing capacity is 0.3B below footing for both the cases.

The ratio of bearing capacity of reinforced soil to that of original soil is termed as bearing capacity ratio (BCR). The BCR values at 10 mm settlement are determined for various cases from Pressure v/s normalized settlement curves and are shown in Figs. 10 and 11 for reinforced and prestressed reinforced sand bed respectively. Figure 12 shows the improvement in BCR when prestressing is applied to geotextile. From

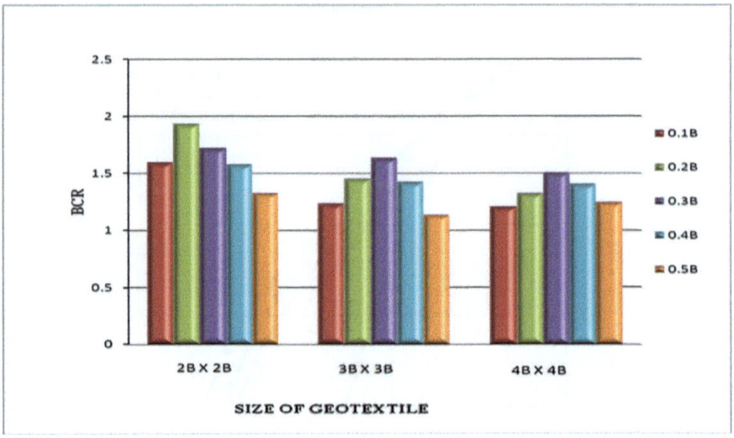

Fig. 10. BCR with geotextile placed at depth 0.1B to 0.5B below footing

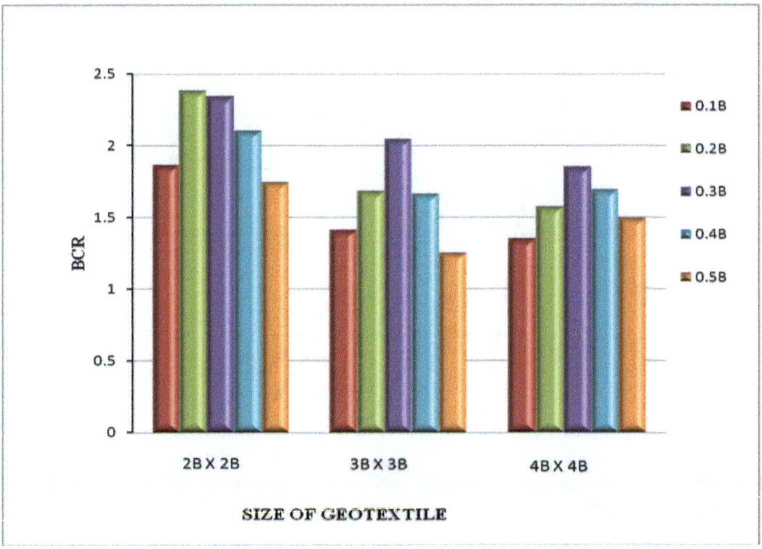

Fig. 11. BCR with prestressed geotextile placed at depth 0.1B to 0.5B below footing

Fig. 12, it is observed that, for all size of geotextile placed at any depth below footing, there is improvement in bearing capacity when it is prestressed.

The ratio of settlement of original soil to that of reinforced soil for same loading is defined as settlement reduction ratio (SRR). For calculation of SRR value, load at 10 mm settlement of original soil is considered as reference load. From Pressure v/s

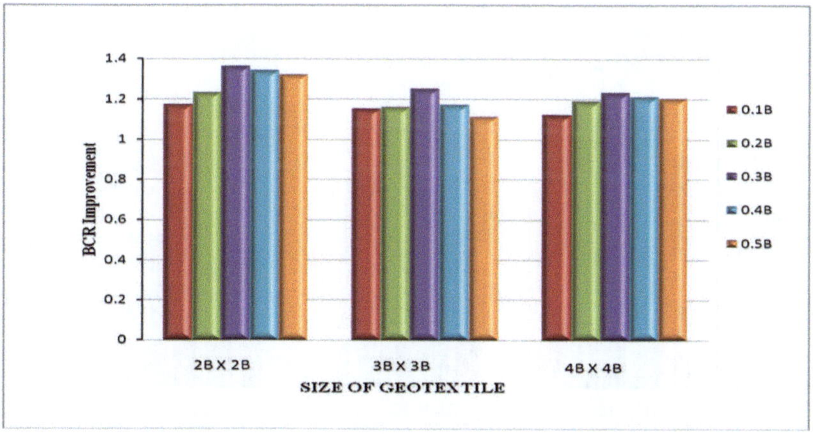

Fig. 12. Improvement in BCR with prestressed geotextile

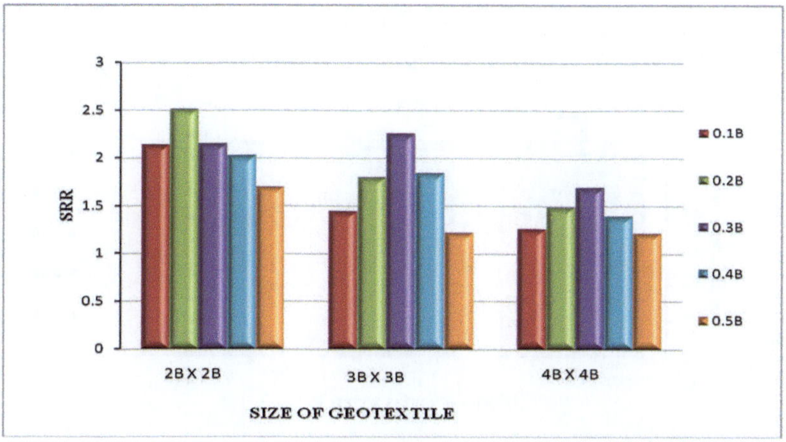

Fig. 13. SRR with geotextile placed at depth 0.1B to 0.5B below footing

normalized settlement curves, SRR values are calculated for various cases and are shown in Figs. 13 and 14 for reinforced and prestressed reinforced sand bed respectively. Figure 15 shows the improvement in SRR when prestressing is applied to geotextile. From Fig. 15, it is observed that, for all size of geotextile placed at any depth below footing, there is reduction in settlement when it is prestressed.

From the present study, the optimum size of geotextile for maximum improvement in bearing capacity and settlement is 2B × 2B and optimum depth of placement of geotextile below footing is 0.2B.

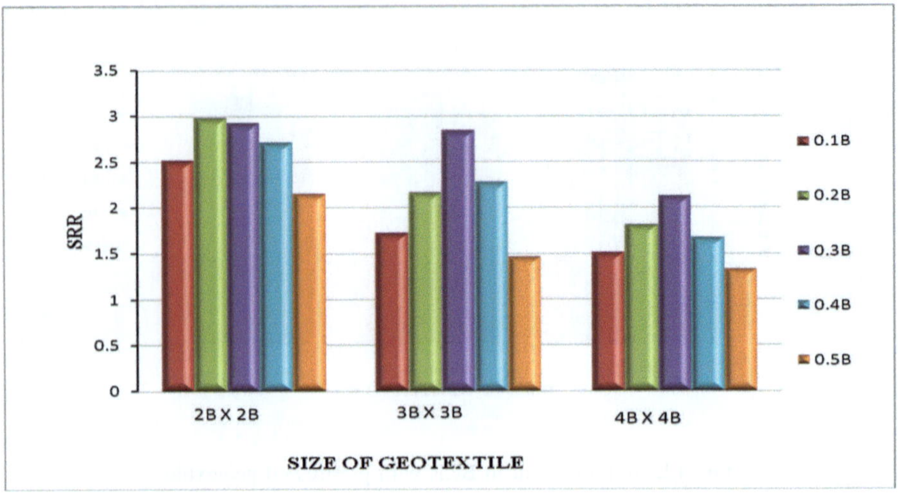

Fig. 14. SRR with prestressed geotextile placed at depth 0.1B to 0.5B below footing

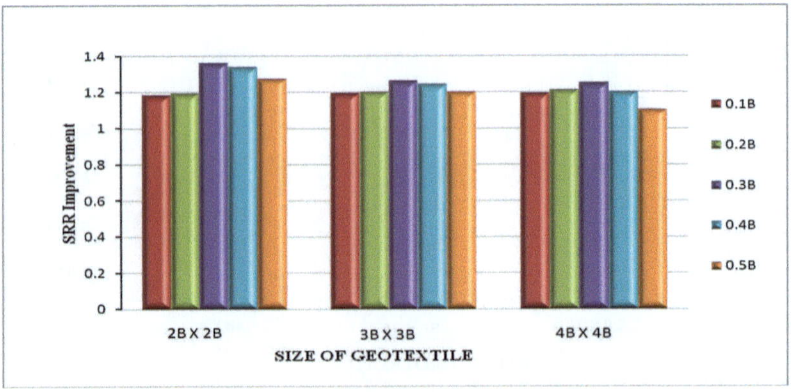

Fig. 15. Improvement in SRR with prestressed geotextile

4 Conclusions

Based on the results obtained from experimental investigation, the following conclusion can be made for geotextile reinforced sand bed. Significant improvements are observed in load bearing capacity and settlement behavior of geotextile reinforced sand bed. The addition of prestress to geotextile gives further improvement. The improvement in bearing capacity depends upon size of geotextile and its placement depth below footing.

References

Adams, M.T., Collin, J.G.: Large model spread footing load tests on geosynthetic reinforced soil foundations. J. Geotech. Geoenviron. Eng. **123**(1), 66–72 (1997)

Basudhar, P.K., Saha, S., Deb, K.: Circular footings resting on geotextile-reinforced sand bed. Geotext. Geomembr. **25**, 377–384 (2007)

Boushehrian, J.H., Hataf, N.: Experimental and numerical investigation of the bearing capacity of model circular and ring footing on reinforced sand. Geotext. Geomembr. **21**, 241–256 (2003)

Guido, V.A., Biesiadecki, G.L., Sullivan, M.J.: Bearing capacity of geotextile reinforced foundation. In: Proceedings of 11th International Conference on Soil Mechanics and Foundation of Engineering, San Francisco, CA, pp. 1777–1780 (1985)

Khing, K.H., Das, B.M., Puri, V.K., Cook, E.E., Yen, S.C.: The bearing capacity of a strip foundation on geogrid reinforced sand. Geotext. Geomembr. **12**, 351–361 (1993)

Lackner, C., Bergado, D.T., Semprich, S.: Prestressed reinforced soil by geosynthethic-concept and experimental investigations. Geotext. Geomembr. **37**, 109–123 (2013)

Latha, G.M., Somwanshi, A.: Bearing capacity of square footing on geosynthetic reinforced sand. Geotext. Geomembr. **27**, 281–294 (2009)

Lovisa, J., Shukla, S.K., Sivakugan, N.: Behaviour of prestressed geotextile-reinforced sand bed supporting a loaded circular footing. Geotext. Geomembr. **28**, 23–32 (2010)

Sitharam, T.G., Sireesh, S.: Model studies of embedded circular footing on geogrid-reinforced sand beds. Ground Improv. **8**(2), 69–75 (2004)

Tafreshi, S.N.M., Dawson, A.R.: A comparison of static and cyclic loading responses of foundations on geocell-reinforced sand. Geotext. Geomembr. **32**(2012), 55–68 (2012)

Yasrobi, S.S., Rahmaninezhad, S.M., Eftekharzadeh, S.F.: Large physical modeling to optimize the geometrical condition of geotextile in reinforced loose sand. In: GeoHunan International Conference 2009, pp. 53–59. ASCE (2009)

Numerical Studies on Ground Improvement Using Geosynthetic Reinforced Sand Layer

G. Sanoop[1(✉)] and Satyajit Patel[2]

[1] Genstru Consultants, Pune, India
sanoopgireshpillai@gmail.com
[2] Applied Mechanics Department, SV National Institute of Technology, Surat 395007, India
spatel@amd.svnit.ac.in

Abstract. Clayey soils exhibit high shrinkage and compressibility characteristics as well as low shear strength. Engineering projects in clayey soils requires construction of deep foundations or use of ground improvement techniques. Soil reinforcement is a popular and widely used ground improvement technique. Shallow foundations resting on geosynthetics reinforced sand layer is a cost effective and feasible construction technique. Since the geosynthetics are placed in sand or granular layer compaction can be easily performed to achieve the design density and adequate friction between sand and the geosynthetics. The performance of strip footings resting on geosynthetics reinforced sand layer overlying clay layer is investigated using finite element software MIDAS GTS NX. A number of numerical models were analyzed and the effect of various parameters such as type of geosynthetic material, depth of sand layer, critical depth of reinforcement below base of footing, number of reinforcement layers, spacing between multiple layer of reinforcement and width of reinforcement layer on the load-settlement behavior of strip footings was studied. The optimum values of these parameters were also determined. Laboratory models of clay underlying sand layer with and without geosynthetics reinforcement were prepared in a steel tank of size 84 * 25 * 50 cm and monotonic load was applied through a steel plate of width 8 cm up to failure. The model test results were compared with the finite element analysis results. Design charts were developed which can be used to determine the depth of sand layer and number of reinforcement layers for a target bearing capacity.

1 Introduction

Naturally occurring clayey soils exhibit high compressibility and low shear strength. Construction of heavy structures on such soils requires erection of deep foundations or implementation of ground improvement techniques such as soil reinforcement. Soil reinforcement or strengthening of soils using geosynthetics or metallic strips have been developed as viable alternative for projects such as retaining walls and embankments for simple and fast construction techniques, better economy, aesthetics, reliability and easily adaptable in variety of environments. Geosynthetics can be classified into categories such as geogrids, geotextiles, geonets and geomembranes based on the methods of manufacturing. Geogrids have an open grid-like appearance and have been used

S.K. Shukla and E. Guler (eds.), *Advances in Reinforced Soil Structures*,
Sustainable Civil Infrastructures, DOI 10.1007/978-3-319-63570-5_8

efficiently to reinforce the soil structures such as embankments, slopes, retaining walls and foundations.

Laman and Yildiz (2003) stated that geogrids generally mobilize a higher soil reinforcement bond stress than geotextiles and have a higher stiffness per weight. Numerous laboratory model test results are currently available in the literature, related to improvement in the load-bearing capacity of shallow foundations supported by sand reinforced with various materials, such as metal strips, metal bars, rope fibres, geotextiles, and geogrids. The results of these investigations clearly showed that the bearing capacity of the foundation can be significantly improved by the inclusion of reinforcement in the ground. Construction of geogrid reinforcement incorporated at the base of a layer of granular fill placed on a soft clay subgrade is commonly used for unpaved roads, embankments, large stabilized areas such as car parks or working platforms for oil drilling and retaining walls Abedi et al. (2009). The use of a geogrid embedded in lightweight granular fill appears to be the most satisfactory means of improving the performance of embankments on very poor foundations. It was shown that reinforcement can significantly reduce the maximum lateral displacements, vertical displacements, and foundation soil heave during embankment construction. Fannin and Sigurdsson (1996) investigated the stabilization of unpaved roads on soft ground with geosynthetics. It was shown that the combination of geosynthetic reinforcement and fill helps to spread the concentrated vertical loads and to inhibit large deformations and local failures. Geosynthetics reinforcing unpaved roads on soft subgrade have been shown to reduce the necessary fill thickness by approximately 30%. Ling and Liu (2001) investigated the performance of geosynthetic-reinforced asphalt pavement under monotonic, cyclic, and dynamic loading conditions. This study showed that geosynthetic reinforcement increased the stiffness and bearing capacity of the asphalt concrete pavement. Under dynamic loading, the life of the asphalt concrete layer was prolonged in the presence of geosynthetic reinforcement. The stiffness of the geogrid and its interlocking with the asphalt concrete contributed to the restraining effect. Zidan (2012) studied the behaviour of circular footing on geogrid-reinforced sand.

In the present study, the bearing capacity and settlement behaviour of strip footings on a sand layer overlaying clayey stratum reinforced with geogrid layers were investigated using finite element software MIDAS GTS NX. The main objectives of this study includes studying the effect of sand layer with and without geosynthetic reinforcement on the settlement behaviour of soft soils, determination of the optimum thickness of sand layer to be provided above clayey soils, determination of the optimum number of geosynthetic layers (geotextiles and geogrids) and the optimum spacing between layers, preparation of design charts to find out required thickness of sand layer and number of geosynthetic layers for a required Safe Bearing Capacity.

2 FEM Modelling

MIDAS GTS NX (version 1.1) is a simulation program developed for the evaluation of soil-structure interaction based on the finite element method. GTS NX helps engineers to perform step-by-step analysis of excavation, banking, structure placement, loading and other factors that directly affect design and construction. The program supports various conditions (soil characteristics, water level etc.) and analytical methodologies to simulate real phenomena. Settings for all types of field conditions can be simulated using non-linear analysis methods (such as linear/non-linear static analysis, linear/non-linear dynamic analysis, seepage and consolidation analysis, slope safety analysis) and various coupled analysis (such as seepage-stress, stress-slope, seepage-slope and nonlinear dynamic-slope coupled analysis).

The first step of modeling involves fixing the dimensions of the test model. The two main factors to be considered for this step is the minimum width of geosynthetics layer required for soil reinforcement and the depth of pressure bulb when the footing is under load. In this study we are considering three various size of footings: 100 cm, 150 cm and 200 cm. So the width of model should be large enough to accommodate the geosynthetics layer for 200 cm footing. The reinforcement effect of geosynthetics is effective in sand layer and very less in cohesive soils. Therefore, a number of models were created and analyzed in GTS NX with sand layer on top and clay at bottom. Geosynthetics of different widths were embedded in the clay layer to find out the minimum width of reinforcement required. Medium clay with properties similar to the clay found at SVNIT Campus was created in GTS NX. Table 1 illustrates the properties of clay and sand bed used in the numerical modeling.

Table 1. Properties of soil

Property	Clay	Sand
Cohesion, *C*	0	0
Angle of internal friction, *Ø*	200	360
Dilatancy angle, *ψ*	0	100
Unit weight, *Y*	1.6 kN/m^3	1.8 kN/m^3
Poisson's ratio, *ν*	0.35	0.3
Modulus of elasticity, *E*	30 MPa	80 MPa
Soil classification	CL	SP

Bearing Capacity Ratio (BCR) is the ratio of bearing capacity of unreinforced soil to the reinforced soil. Width ratio is the ratio of width of geosynthetics to the width of footing. Since BCR depends on width of reinforcement, it is increased till it reaches such a point that there is no further increase in BCR with increase in width. From Fig. 1, BCR reaches maximum at width ratio of geosynthetics layer equal to 6. Therefore, the minimum width of reinforcement required for 200 cm footing is 12 m. Hence clay model with dimensions 14 m * 14 m * 14 m was created in GTS NX with medium Hybrid mesher. Mohr – Coulomb model was chosen for the model. Load was applied similar to the plate load test used in model testing in laboratory. Figure 2 shows a general schematic representation of the clay-sand-geosynthetic model used for analysis and the parameters that are studied and Fig. 3 shows 3D stress contour in MIDAS GTS NX.

Since the dimensions of geosynthetics, footing and soil model is fixed, parametric studies are conducted using GTS NX. Firstly, model consisting of only clay is analysed and then top layer of clay is replaced with varying thickness of sand till the optimum

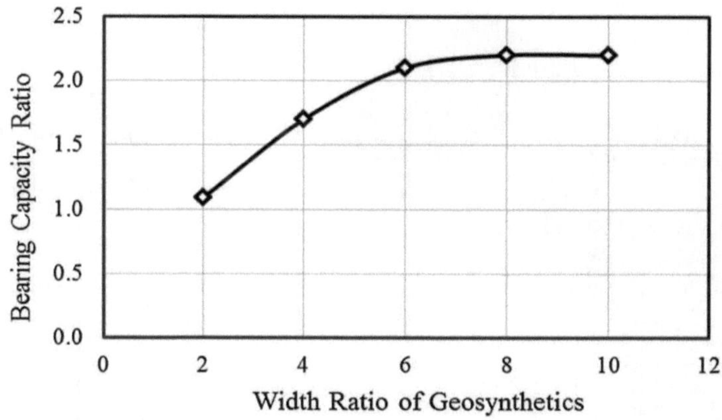

Fig. 1. Relationship between BCR and width ratio (b/B) of geosynthetic layer

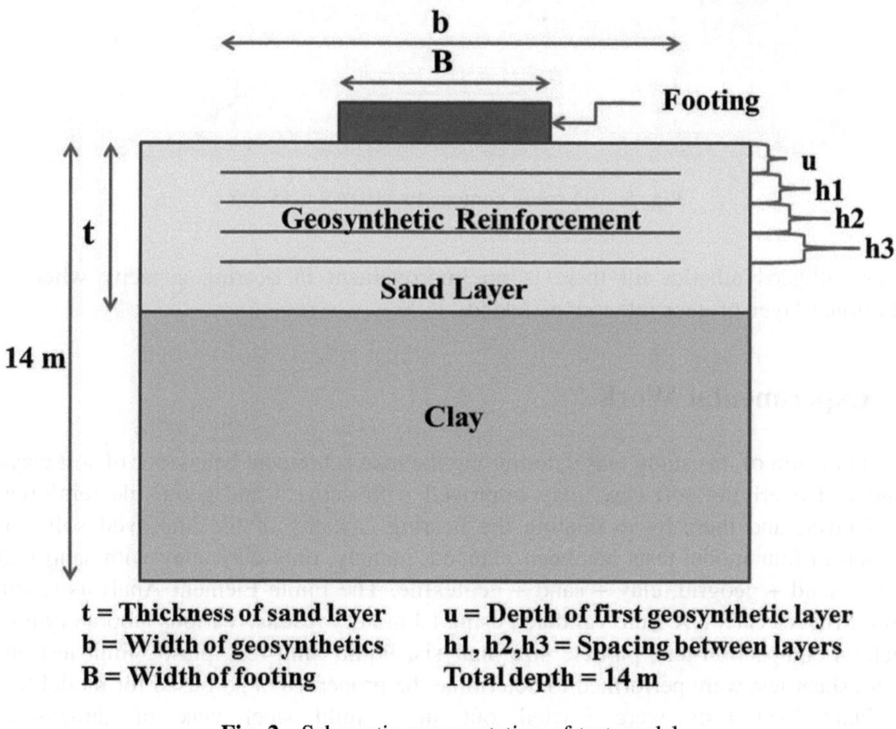

Fig. 2. Schematic representation of test model

thickness of sand is reached. After determining the optimum thickness of sand layer, the depth of first layer of geosynthetics is finalised by varying the depth of geosynthetics layer with respect to the width of footing. When depth of first layer is fixed, same procedure is repeated for second layer. This procedure is carried out for multiple

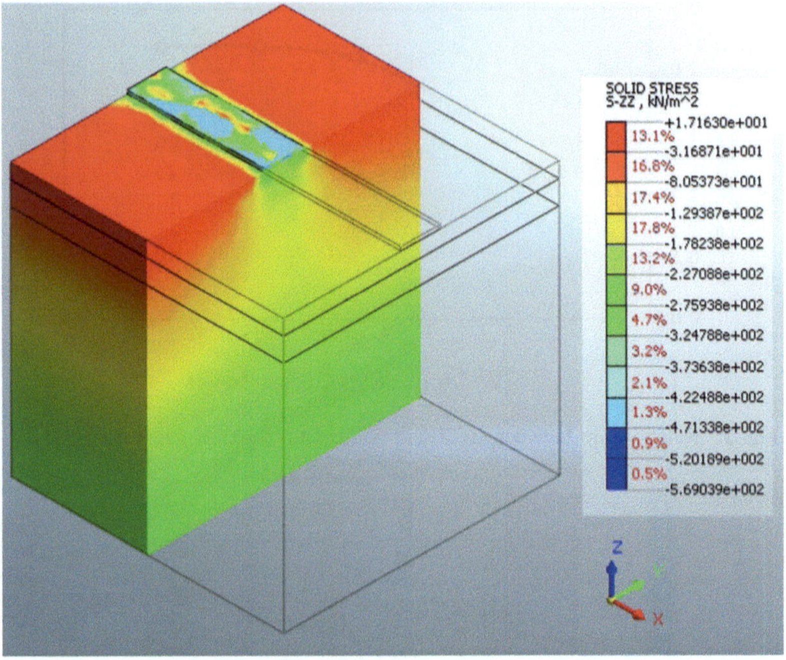

Fig. 3. 3D stress contour in MIDAS GTS NX

layers of geosynthetics till there is no improvement in bearing capacity when an additional layer of geosynthetics is added.

3 Experimental Work

The main aim of this study was determining the load settlement behaviour of soft clays, sand bed overlying soft clay, clay improved with geogrid and geotextile reinforced sand layer, and there by evaluating the bearing capacity of the improved soft soil. A total of four model tests has been planned, namely, only clay, clay with sand bed, clay + sand + geogrid, clay + sand + geotextile. The Finite Element Analysis results from MIDAS GTS NX were validated using laboratory results. Various laboratory tests such as compaction test, particle size analysis, liquid limit test, plastic limit test and direct shear test were performed to determine the properties of soil used for model test.

Plate load tests were carried out in a mild steel tank of dimensions 84 * 25 * 50 cm. A mild steel loading plate of dimensions 25 * 8 * 2 cm was used as footing to transfer load to soil mass. Load was applied using hydraulic jack. To simulate ground condition of overlying soil above footing, surcharge load equal to 20 cm deep soil mass was applied on the soil surface. Two dial gauges were attached to the loading plate to measure deflections of plate while loading.

4 Results and Discussion

4.1 Experimental Work

A total of four sets of plate load tests were carried out in laboratory and the Applied pressure vs Settlement relationships were obtained. Mild steel plate of dimensions 8 * 25 * 2 cm was used as the strip footing to transfer load from hydraulic jack to the surface of soil. Surcharge load equal to 20 cm of soil layer was applied on the exposed surface of soil. Clay failed at 320 kPa stress, unreinforced soil failed at 400 kPa, whereas geogrid and geotextile reinforced soils achieved bearing capacity of 730 kPa and 940 kPa respectively. That is, the Bearing Capacity Ratio (BCR) of geotextile reinforcement is 1.825 and that of geogrid reinforcement is 2.35.

The results obtained from laboratory experiments were validated in FEM using GTS NX software. Figure 4 shows the results of FEM validation of laboratory results. The results are nearly matching for all three cases with a slight deviation. The applied pressure is lesser for same settlements in case of FEM results till elastic limits. This may be due to various reasons. Reinforcement by geogrid is mainly from lateral constraint provided by interlocking between aggregates and geogrid. But geotextile functions through a number of ways, including reinforcement through interaction friction, separation between subgrade soil and base course material, filtration, and drainage. In FEM analysis using software, these factors may not be considered for the reinforcement mechanism of geosynthetics. However, the results obtained from the experiment and the FEM analysis match to a great extent and therefore results are acceptable.

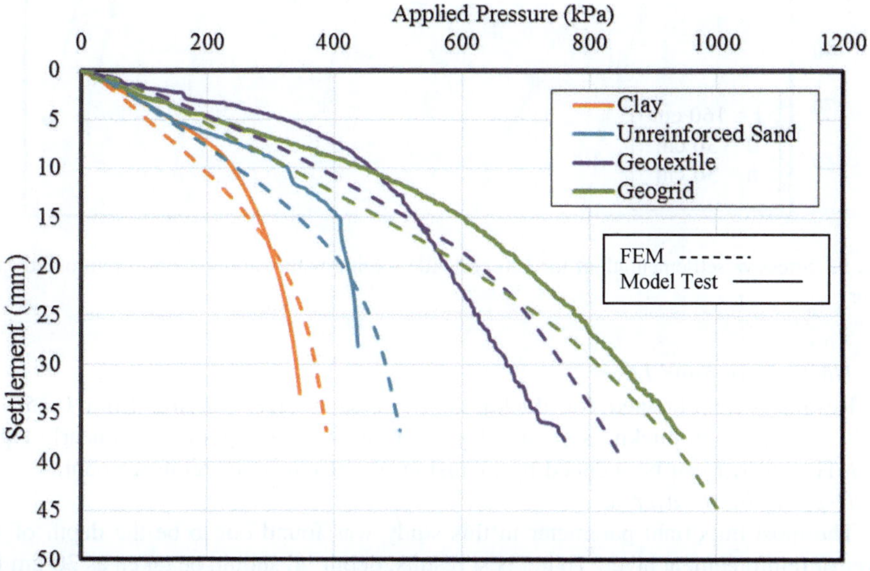

Fig. 4. Comparison of FEM and experimental results for geosynthetic-reinforced soils

4.2 FEM Analysis

A series of numerical analyses were carried out in medium clay models with and without ground improvement using geosynthetics reinforced sand layer. Models with only clay, sand overlying clay bed, and sand overlying clay reinforced with multiple number of geogrids and geotextiles were analysed in FEM software MIDAS GTS NX. Parameters varied in this study are thickness of sand layer, number of geosynthetics layers, spacing between layers and depth of first layer of reinforcement below base of footing. Parameters kept unchanged in this study are total depth of soil model and properties of clay and sand.

Three footing sizes, namely 100 cm, 150 cm and 200 cm were selected for both geogrid reinforced soils and geotextile reinforced soils. Figures 5 and 6 shows the stress vs settlement relationship for 100 cm footings with geogrids and geotextiles respectively.

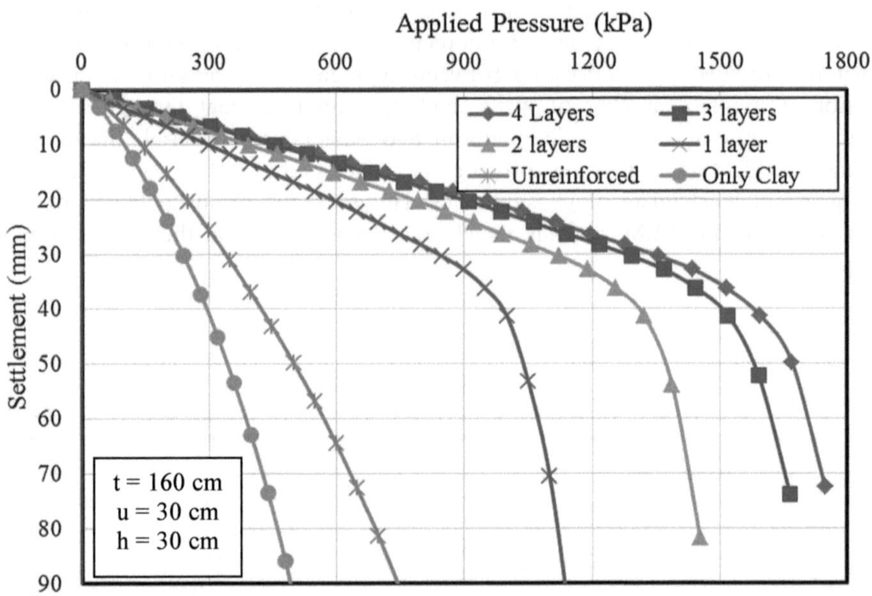

Fig. 5. Stress vs settlement chart for 100 cm wide footing with/without geogrid reinforced sand layer

Thickness of Sand Layer 't'

Results were obtained for thickness 't' of sand layer varying from $0.25B$ to $1.5B$. The optimum thickness of sand layer for varying footing sizes was nearly equal to B. However, t can be reduced by inclusion of geosynthetics reinforcement.

Depth of First Layer 'u'

The most important parameter in this study was found out to be the depth of 1st layer of reinforcement layer. To get best results, depth 'u' should be taken as 30 cm for 100 cm footing, 40 cm for 150 cm footing and 45 cm for 200 cm footing.

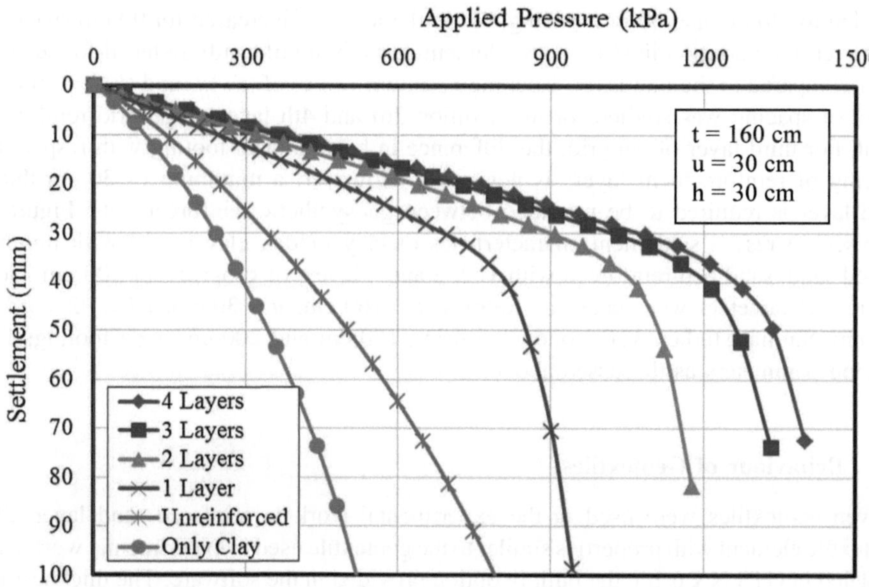

Fig. 6. Stress vs settlement chart for 100 cm wide footing with/without geotextile reinforced sand layer

Spacing Between Layers 'h'

Spacing between 1st to *n*th layer of geosynthetics can vary between 30 cm to 50 cm, but best results are obtained by selecting '*h*' as 30 cm.

Number of Reinforcement Layers 'N'

Maximum improvement in bearing capacity was obtained when the first layer of reinforcement is added. With addition of each layer of reinforcement, bearing capacity increases and settlements are reduced. However, there was no improvement observed after 4 layers of geosynthetics layers.

4.3 Behaviour of Geogrids

Clay model loaded with 100 cm wide footing was analyzed with varying thickness of sand layer below base of the footing. Thickness of sand layer was varied from 40 cm to 160 cm. It was observed that the bearing capacity of model increases with increase in value of '*t*', but the difference in bearing capacity was reduced when '*t*' is increased beyond 120 cm. Therefore 160 cm thick sand layer is selected for the addition of geogrid reinforcement. Thickness of sand bed was taken as 160 cm and the depth '*u*' was varied between 30 cm and 90 cm. For values of '*u*' between 30 cm and 60 cm, the difference in stress vs settlement behaviour is low. But on further increase of '*u*', higher settlements are observed for the same loading. First layer of geogrid was placed at 30 cm from bottom face of footing for analysis of multiple layers of geogrids. Sand bed '*t*' was taken as 160 cm and '*u*' equal to 30 cm. Second layer of geogrid was placed at spacing 30 cm to 90 cm. The load- settlement characteristics are not significantly affected for 30 cm and 60 cm spacing, but settlements are increased for 90 cm spacing. However, the variation in stress vs settlement curve is significantly reduced for second layer compared to the first layer. Keeping optimum values of '*t*', '*u*' and '*h1*' = 30 cm, effect of spacing was studied for the addition 3rd and 4th layer of geogrid reinforcement. For third layer of geogrid, the difference in behaviour of footing with respect to spacing of reinforcement layers is negligible. However, a minimum of 30 cm thick sand layer is required to be provided between geosynthetic reinforcements. Figure 5 represents stress – settlement characteristics of clay model, clay model with unreinforced sand, sand bed reinforced with 1, 2, 3 and 4 layers of geogrids for 100 cm wide footing. Parameters were taken as follows: t = 160 cm, u = 30 cm, $h1$ = $h2$ = $h3$ = 30 cm. Similar studies were conducted using 150 cm and 200 cm wide footings for varying parameters as discussed above.

4.4 Behaviour of Geotextiles

Woven geotextiles were used in the experimental work to reinforce sand layer. 2D geotextile element with properties similar to the geotextile used in experimental work was modeled in GTS NX using the built in option provided in the software. The thickness of geotextile was taken as 1.5 mm. The performance of woven geotextiles in FEM analysis is slightly lesser compared to experimental work. This may be due to the assumptions in FEM which fails to model the interaction friction of geotextile layer accurately. The effect of first layer of geotextiles and the spacing between multiple layers were found similar to that of geogrids, that is, '*u*' equal to 0.3*B* to 0.6*B* and '*h*' equal to 0.3*B* to 0.4*B*. Figure 6 represents stress – settlement characteristics of clay model, clay model with unreinforced sand, sand bed reinforced with 1, 2, 3 and 4 layers of geogrids for 100 cm wide footing. We got optimum thickness of sand layer for 100 cm footing as 120 cm. Similar to geogrids depth '*u*' and spacings '*h1*', *h2*' and '*h3*' was varied by placing geotextiles at various depths and they were analyzed in GTS NX. Parameters were taken as follows: t = 120 cm, u = 300 cm, $h1$ = $h2$ = $h3$ = 30 cm. Similar studies were conducted using 150 cm and 200 cm wide footings for varying parameters as discussed above.

4.5 Bearing Capacity Ratio

Bearing Capacity Ratio is defined as the ratio of ultimate bearing capacity of unreinforced soils to that of reinforced soils, as described in Eq. (1). Figure 7 shows relationship of BCR and number of reinforcement layers for geogrids. On addition of first layer of geogrid, BCR of 2.2 was achieved and the maximum BCR achieved was 3.6, by adding 4 layers of geogrid. Addition of geogrid layer after 4th layer does not show any significant increase in bearing capacity or reduction in settlement.

$$\mathrm{BCR} = q_r/q_u \tag{1}$$

where

q_r = Bearing capacity of reinforced sand,
q_u = Bearing capacity of unreinforced sand.

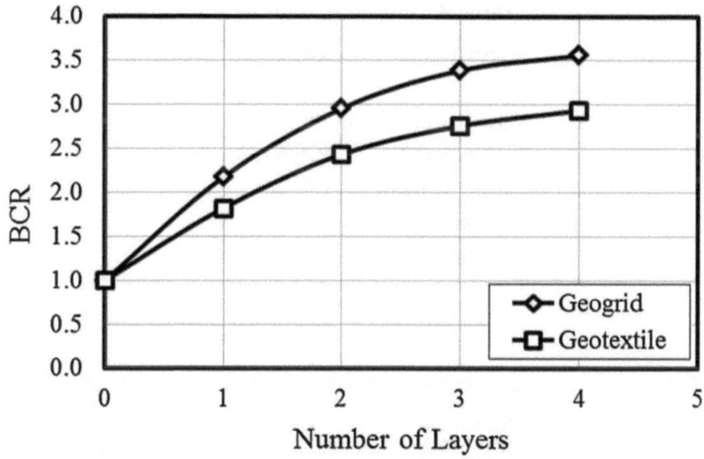

Fig. 7. Comparison of BCR of geogrids and geotextiles for 150 cm wide footing

On addition of first layer of geotextiles, BCR of 1.8 was achieved and the maximum BCR achieved was 2.9, by adding 4 layers of geogrid. Optimum number of layers was found to be 4 for geotextiles. Figure 7 compares the performance of geogrids and geotextiles in improving bearing capacity under similar conditions.

From Fig. 7, it is observed that geogrids exhibit superior performance compared with geotextile when it is used for soil reinforcement. Although expensive, geogrids give higher bearing capacity when used for foundation purposes. But in field conditions, geotextiles have versatile uses in addition to soil reinforcement such as separation, filtration, drainage etc. These factors also indirectly affect bearing capacity of soils. For example, preventing subgrade soil mixing with base material is crucial for the strength of pavements. Therefore, selection of reinforcing material not only depends on target bearing capacity, but also the site conditions and purpose of construction. Table 4 shows the percentage increase in Bearing Capacity for different number of geogrids and geotextiles.

Table 2. Properties of geogrid

Property	Value
Structure	Uniaxial
Aperture shape	Rectangular
Aperture size	51 mm × 31 mm
Mass per unit area	500 g/m^2
Raw material	Polypropylene
Elastic modulus	2 GPa
Poisson's ratio	0.3
Thickness	0.003 m
Elongation at nominal strength	8%
Tensile strength	80 kN/m

Table 3. Properties of geotextile

Property	Value
Type	Woven
Mass per unit area	300 g/m^2
Raw material	Polypropylene
Elastic modulus	1.5 GPa
Poisson's ratio	0.3
Thickness	0.0015 m
Tensile strength	45 kN/m

Table 4. Percentage increase in Bearing Capacity w.r.t only clay layer

Reinforcement	Geogrid	Geotextile
Only clay	-	-
Unreinforced sand	5.76%	5.76%
1 Layer	129.5%	92.3%
2 Layers	212.5%	157.2%
3 Layers	258.2%	191.8%
4 Layers	276.2%	210.3%

4.6 Design Charts

Studies were conducted on different parameters such as number of reinforcement layers '*N*', thickness of strip footing '*B*', thickness of sand layer '*t*', spacing between layers '*h*' and depth of reinforcement below base of footing '*u*' with respect to Bearing Capacity. Based on the findings from these studies, we can choose required *u*, *t*, *N* and *h* for target Safe Bearing Capacity and economy. For field applications, design charts are prepared considering thickness of sand layer, number of geosynthetic layers and Safe Bearing Capacity. Since depth of reinforcement layer below footing and spacing between layers can be optimized, optimum '*u*' and '*h*' is chosen for preparation of design charts. Therefore, for a target Safe Bearing Capacity, we can obtain required sand thickness and number of reinforcement layers (under specified depth of reinforcement and spacing between layers). These design charts is valid only for medium clays (clays which can be indented with strong thumb pressure). Figure 8 shows design chart for 100 cm wide footings on geogrid reinforced sand layer. Depth of first geogrid layer below base of footing was taken as 30 cm and the spacing between adjacent geogrid layers was 30 cm. Figure 8 shows the variation of Safe Bearing Capacity (SBC) with respect to thickness of sand bed '*t*'. Five different curves on the graph represents the behaviour of footings on unreinforced soils, soils reinforced with 1 layer, 2 layers, 3 layers and 4 layers of geogrids.

Similar to the design charts for geogrids, the same was prepared for footings resting on geotextile reinforced sand layer. Figure 9 shows design chart for 100 cm wide footings on geotextile reinforced sand layer. Depth of first geogrid layer below base of footing was taken as 30 cm and the spacing between adjacent geogrid layers was

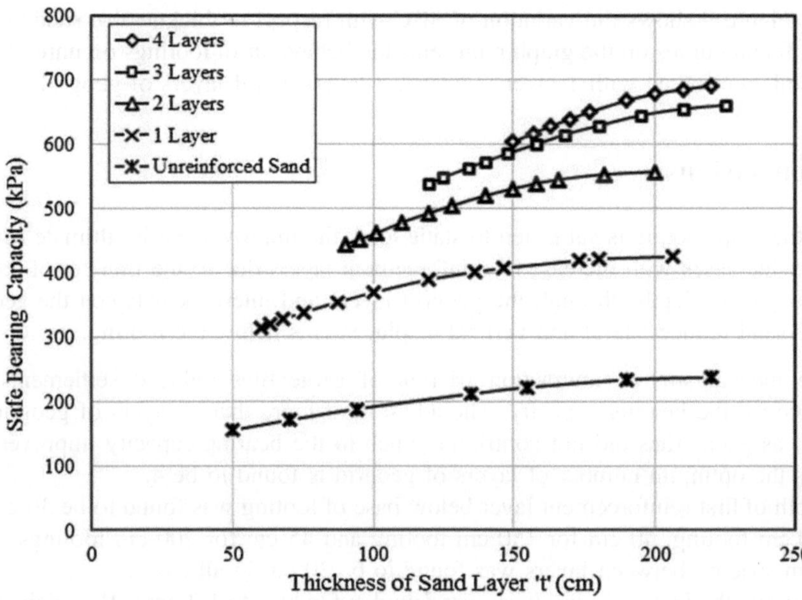

Fig. 8. Design chart for geogrid - reinforced sand layer for 100 cm wide footing

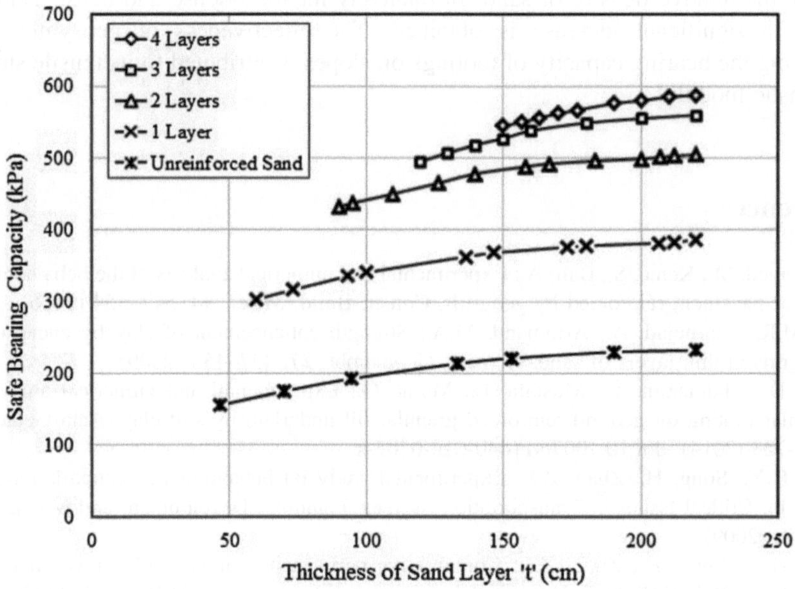

Fig. 9. Design chart for geotextile - reinforced sand layer for 100 cm wide footing

30 cm. Figure 9 shows the variation of SBC with respect to thickness of sand bed '*t*'. Five different curves on the graph represents the behaviour of footings on unreinforced soils, soils reinforced with 1 layer, 2 layers, 3 layers and 4 layers of geogrids.

5 Conclusions

When the strip footing is subjected to static load, the improvement in ultimate bearing capacity increases with increase of reinforcement layers due to the transfer of footing loads to greater depths through the geogrid layers and interlock between the geogrid and the sand reduce lateral and vertical displacements below the footing.

- The mechanisms of interaction friction of geotextiles reduced settlements and improved the bearing capacity. The addition of more than 4 layers of geogrids as well as geotextiles did not contribute much to the bearing capacity improvement: thus the optimum number of layers of geogrid is found to be 4.
- Depth of first reinforcement layer below base of footing was found to be 30 cm for 100 cm footing, 40 cm for 150 cm footing and 45 cm for 200 cm footings. Optimum spacing between layers was found to be 30 cm in all cases.
- Optimum thickness of sand layer was found out to be 1 to 1.2 times B, which can be reduced by inclusion of geosynthetics layers.

The improvement in bearing capacity using geosynthetic reinforcement is dependent on the relative density of sand. In relatively medium-dense and dense sand conditions, a significant increase is obtained. The effectiveness of geosynthetics in improving the bearing capacity of footings on slopes is attributed to its tensile strength and elastic modulus.

References

Abdessemed, M., Kenai, S., Bali, A.: Experimental and numerical analysis of the behaviour of an airport pavement reinforced by geogrids. Constr. Build. Mater. **94**, 547–554 (2015)

Abdi, M.R., Sadrnejad, A., Arjomand, M.A.: Strength enhancement of clay by encapsulating geogrids in thin layers of sand. Geotext. Geomembr. **27**, 447–455 (2009)

Ahmet, D., Abdulazim, Y., Mustafa, L., Murat, O.: Experimental and numerical analyses of circular footing on geogrid-reinforced granular fill underlain by soft clay. Acta Geotech. **9**, 711–723 (2014). doi:10.1007/s11440-013-0207-x

Chang, H.Y., Song, H., Zhao, Z.J.: Experimental study on behaviour of geotextile-reinforced soil. In: Critical Issues in Transportation Systems Planning, Development and Management. ASCE (2009)

Chang, H.Y., Song, H., Zhou, Z.J.: Compressive performance of geogrid-reinforced granular soil. In: 3126 ICCTP 2010: Integrated Transportation Systems, pp. 3126–3132. ASCE (2010)

Dong, Y.L., Han, J., Bai, X.H.: A numerical study on stress-strain responses of biaxial geogrids under tension at different directions. In: GeoFlorida: Advances in Analysis, Modelling & Design (GSP 199) © ASCE, pp. 2551–2560 (2010)

Fannin, R.J., Sigurdsson, O.: Field observations on stabilization of unpaved roads with geosynthetics. J. Geotech. Eng. **122**(7), 544–553 (1996)

Fathi, A., Khaled, G., Enas, A.O.: Behavior of strip footings on reinforced and unreinforced sand slope. GeoCongress: Geosustainability and Geohazard Mitigation, pp. 25–32. ASCE (2008)

Fei, Y.Y., Yang, Y.H.: Fem analysis on geogrid reinforced asphalt concrete pavement. In: Proceedings of 4th Asian Regional Conference on Geosynthetics, Shanghai, 17–20 June 2008

Ferrotti, G., Caestrari, F., Virgili, A.: A strategic laboratory approach for the performance investigation of geogrids in flexible pavements. Constr. Build. Mater. (25-032) (2010)

Florian, B., Joseph, C.: Recent research and future implications of the actual behaviour of geogrids in reinforced soil. In: Earth Retention Conference (ER 2010), pp. 460–477. ASCE (2010)

Hao, W., Baoshan, H., Zixin, Z.: Evaluation of the Reinforcement effect of geogrids in pavement base using loaded wheel tester (LWT). In: GeoShanghai International Conference, pp. 314–319. ASCE (2010)

Hiroshan, H., Louis, G.: Use of geogrids to enhance stability of slope in bioreactor landfills: a conceptual method. In: International Foundation Congress and Equipment Expo, pp. 520–526 (2009)

Jie, H., Ken, A.: Use of geogrid reinforced and pile supported earth structures. In: Deep Foundations, pp. 668–678. ASCE (2002)

Lai, J., Chung, S.Y., Yang, B.H., Wu, S., Pan, C.L.: Numerical study on enhancing the bearing capacity of shallow foundation using geosynthetics. In: Innovative and Sustainable Use of Geomaterials and Geosystems, GSP 245, pp. 64–70. ASCE (2014)

Laman, M., Yildiz, A.: Model studies of ring foundations on geogrid-reinforced sand. Geosynth. Int. **10**(5), 142–152 (2003)

Ling, H.I., Liu, Z.: Performance of geosynthetic reinforced asphalt pavements. J. Geotech. Geoenviron. Eng. **127**(2), 177–184 (2001)

Khing, K.H, Das, B.M, Puff, V.K, Cook, E.E., Yen, S.C.: The bearing-capacity of a strip foundation on geogrid-reinforced sand. Geotext. Geomembr. 0266–1144 (2003). Elsevier

Mane, A.S., Viswanadham, B.V.S.: Studies on the performance of geogrid reinforced soil walls with compressible inclusion. In: GeoCongress, pp. 1340–1349. ASCE (2012)

Marko, C., Krunoslav, M., Sanja, D.: Influence of reinforcing geogrids on soil properties. Tehnički Vjesnik **13**(3,4), 21–25 (2006)

Maubeuge, K.V., Klompmaker, J.: New developments for geogrid reinforced base courses. In: Geo-Frontiers, pp. 4624–2634. ASCE (2011)

McGown, A., Kupec, J., Maubeuge, V.: Testing biaxial geogrids for specification and design purposes. In: GRI-18 Geosynthetics Research and Development in Progress (2005)

Nelson, L., Fountain, G.B., Wayne, M.H.: Performance verification of a geogrid mechanically stabilized layer flexible pavement design as part of the La Media Road widening project. In: GeoCongress, pp. 1391–1399. ASCE (2012)

Ngoc, T.N., Indraratna, B., Cholachat, R.: DEM simulation of the behaviour of geogrid stabilised ballast fouled with coal. Comput. Geotech. **55**, 224–231 (2014)

Nicola, M., Giuseppe, C.: Deformative behaviour of different geogrids embedded in a granular soil under monotonic and cyclic pullout loads. Geotext. Geomembr. **32**, 104–110 (2011)

Ragui, F., Koerner, R., Sansone, L.: Experimental behavior of polymeric geogrids in pullout. J. Geotech. Eng. **120**, 661–677 (2008). ASCE

Reza, N., Ebrahim, M.: Bearing capacity of two close strip footings on soft clay reinforced with geotextile. Arab. J. Geosci. **7**, 623–639 (2014)

Salih, K., Mustafa, L.: Experimental and numerical studies of strip footings on geogrid-reinforced sand slope. Arab. J. Sci. Eng. (2014). doi:10.1007/s13369-013-0795-7

Saran, S.: Reinforced Soils and Its Engineering Applications, pp. 201–259. IKI Publications, New Delhi (2013)

Syed, K.K.H., Indraratna, B., Jayan, S.V.: Performance assessment of geogrid-reinforced railroad ballast during cyclic loading. Transp. Geotech. **2**, 99–107 (2014)

Tanchaisawat, T., Bergado, D., Lai, Y.P.: Numerical simulation of geogrid reinforced lightweight geomaterials on soft ground area. In: Proceedings of 4th Asian Regional Conference on Geosynthetics, Shanghai, 17–20 June (2008)

Xiaochao, T., Shelley, S., Palomino, A.M.: Mechanistic-empirical performance prediction of geogrid-modified soft soil subgrade. In: Geo-Congress 2014 Technical Papers, GSP 234, pp. 3054–3061. ASCE (2014)

Xiaohui, S., Jie, H., Jayhyun, K., Parsons, R.L.: Radial stresses and resilient deformations of geogrid-stabilized unpaved roads under cyclic plate loading tests. Geotext. Geomembr. **4**, 440–449 (2015)

Yeo, S., Hsuan, Y.G.: Evaluation of creep behaviour of high density polyethylene and polyethylene-terephthalate geogrids. Geotext. Geomembr. **28**, 409–421 (2010)

Zidan, A.F.: Numerical study of behaviour of circular footing on geogrid-reinforced sand under static and dynamic loading. Geotech. Geol. Eng. **30**, 499–510 (2012). doi:10.1007/s10706-011-9483-0

Zhang, J., Hurta, G.: Comparison of geotexile and geogrid reinforcement on unpaved road. In: Geocongress: Geosustainability and Geohazard Mitigation. ASCE (2008)

Bearing Capacity Prediction of Inclined Loaded Strip Footing on Reinforced Sand by ANN

R. Sahu[1(✉)], C.R. Patra[1], N. Sivakugan[2], and B.M. Das[3]

[1] National Institute of Technology Rourkela, Rourkela, India
roma.sahu.civ@gmail.com, crpatra19@yahoo.co.in
[2] James Cook University, Townsville, Australia
siva.sivakugan@jcu.edu.in
[3] California State University, Sacramento, USA
brajamdas@gmail.com

Abstract. Laboratory model tests have been conducted on a strip foundation resting over multi-layered geogrid-reinforced dense and loose sand subjected to inclined load. Based on the laboratory model test results, a neural network model is developed to estimate the reduction factor for bearing capacity. The reduction factor obtained by ANN can be used to estimate the ultimate bearing capacity of a strip foundation subjected to centric inclined load from the ultimate bearing capacity of the same foundation under centric vertical loading. A thorough sensitivity analysis was carried out to find out the important parameters affecting the reduction factor. Emphasis was given on the construction of neural interpretation diagram, based on the weights developed in the neural network model, to determine the direct or inverse effect of input parameters to the output. An ANN model equation is developed based on trained weights of the neural network model. The results from artificial neural network (ANN) were compared with the laboratory model test results and these results are in good agreement.

Keywords: Inclined load · Geogrid · Sand · Neural network · Ultimate bearing capacity · Reduction factor

1 Introduction

During the last thirty years, a number of laboratory model test results and a few field test results have been published that are related to the ultimate bearing capacity of shallow foundation resting over geogrid reinforced sand and clay. Most of the experimental studies were related to centric loading condition. However, none of the published studies address the effect of load inclination on the ultimate bearing capacity of strip foundation resting over multi-layered geogrid reinforced soil. The purpose of this study is to develop a neural network model from the results of laboratory model tests conducted by Sahu *et al.* (2016) to estimate the reduction factor, *RF*. This *RF* is the ratio of the ultimate bearing capacity of strip foundation on reinforced soil subjected to an inclined load to the ultimate bearing capacity of footing subjected to a

S.K. Shukla and E. Guler (eds.), *Advances in Reinforced Soil Structures*,
Sustainable Civil Infrastructures, DOI 10.1007/978-3-319-63570-5_9

centric vertical load at the same depth of embedment. In the present study, a feed-forward back-propagation neural network is trained with Levenberg-Marquardt algorithm. A thorough sensitivity analysis is made to interpret the important input variables. Neural Interpretation diagram is constructed based on the weights of the developed neural network model, to determine whether the input parameters have direct or inverse effect on the output. A prediction model equation is developed based on the weights of the neural network model. The predicted reduction factor is compared with the empirical equation proposed by Sahu *et al.* (2016).

2 Database and Preprocessing

The extensive database of laboratory model test results available in Sahu *et al.* (2016) has been employed in the present study. Load tests were carried out on model strip foundation resting on geogrid reinforced sand subjected to inclined loads as shown in Fig. 1. The details of the tests and the procedure have been described in Sahu *et al.* (2016). The database used in the present analysis is presented in Table 1. The database consist of parameters like load inclination α, embedment ratio D_f/B, depth of reinforcement measured from the bottom of the foundation d, friction angle ϕ and ultimate bearing capacity q_{uR}. Eighty laboratory model tests were conducted. α/ϕ and d_f/B are used as the two dimensionless input parameters in the ANN model and the output is the reduction factor (RF). The reduction factor (RF) is given by

$$RF = \frac{q_{uR\left(\alpha/\varphi,\, d_f/B\right)}}{q_{uR\left(\alpha/\varphi=0,\, d_f/B\right)}} \tag{1}$$

where $q_{uR\left(\alpha/\varphi,\, d_f/B\right)}$ is the ultimate bearing capacity with load inclination ratio of α/φ at an normalized depth of reinforcement layer ratio d_f/B and $q_{uR\left(\alpha/\varphi=0,\, d_f/B\right)}$ is the ultimate bearing capacity with centric vertical loading (i.e. $\alpha/\phi = 0$) at depth of

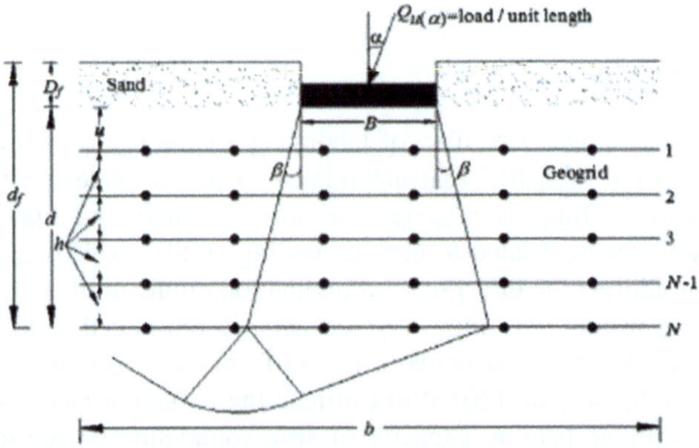

Fig. 1. Strip foundation over geogrid-reinforced soil subjected to inclined ultimate load

Table 1. Database used for ANN model and comparison with Sahu *et al.* (2016)

Data type (1)	Test no. (2)	d_f/B (3)	(α/ϕ) (4)	Experimental q_u (kN/m^2) (5)	RF (expt) (6)	RF (ANN) (7)	RF (Pred) (8)
Training	1	0.350	0.122	208	0.754	0.754	0.752
	2	0.350	0.244	159	0.576	0.595	0.578
	3	0.350	0.367	116	0.420	0.415	0.423
	4	0.350	0.489	74	0.268	0.258	0.279
	5	0.600	0.000	370	1.000	1.000	1.000
	6	0.600	0.122	272	0.735	0.719	0.741
	7	0.600	0.244	208	0.562	0.565	0.559
	8	0.600	0.367	145	0.392	0.401	0.397
	9	0.850	0.000	550	1.000	0.999	1.000
	10	0.850	0.122	400	0.727	0.708	0.734
	11	0.850	0.244	300	0.545	0.537	0.546
	12	0.850	0.489	123	0.224	0.233	0.226
	13	1.100	0.000	640	1.000	1	1.000
	14	1.100	0.122	456	0.713	0.716	0.728
	15	1.100	0.367	230	0.359	0.374	0.367
	16	1.100	0.489	134	0.209	0.221	0.210
	17	1.35	0.00	500	1.000	1.013	1.000
	18	1.35	0.24	385	0.770	0.774	0.764
	19	1.35	0.37	317	0.634	0.63	0.653
	20	1.35	0.49	250	0.500	0.481	0.529
	21	1.6	0.122	625	0.887	0.918	0.885
	22	1.6	0.244	528	0.749	0.755	0.761
	23	1.6	0.367	430	0.610	0.620	0.633
	24	1.6	0.489	333	0.472	0.471	0.502
	25	1.85	0.000	820	1.000	1.005	1.000
	26	1.85	0.122	725	0.884	0.889	0.880
	27	1.85	0.244	608	0.741	0.743	0.750
	28	1.85	0.367	490	0.598	0.607	0.615
	29	2.1	0.000	930	1.000	0.998	1.000
	30	2.1	0.122	810	0.871	0.860	0.875
	31	2.1	0.244	675	0.726	0.732	0.739
	32	2.1	0.489	382	0.411	0.451	0.455
	33	0.35	0.000	85	1.000	1.000	1.000
	34	0.35	0.147	63	0.741	0.714	0.714
	35	0.35	0.441	28	0.329	0.320	0.334
	36	0.35	0.588	13	0.153	0.155	0.169
	37	0.6	0.000	115	1.000	1.000	1.000
	38	0.60	0.29	58	0.504	0.495	0.491
	39	0.60	0.44	35	0.304	0.308	0.305
	40	0.60	0.59	16	0.139	0.146	0.133
	41	0.85	0.15	101	0.697	0.682	0.693
	42	0.85	0.29	70	0.483	0.477	0.477
	43	0.85	0.44	41	0.283	0.294	0.285

(*continued*)

Table 1. *(continued)*

Data type (1)	Test no. (2)	d_f/ B (3)	(α/ϕ) (4)	Experimental q_u (kN/m^2) (5)	RF (expt) (6)	RF (ANN) (7)	RF (Pred) (8)
	44	0.85	0.59	19	0.131	0.138	0.108
	45	1.10	0.00	178	1.000	1	1.000
	46	1.10	0.15	121	0.680	0.673	0.687
	47	1.10	0.29	82	0.461	0.463	0.466
	48	1.10	0.44	47	0.264	0.281	0.270
	49	1.35	0.00	118	1.000	1.013	1.000
	50	1.35	0.15	106	0.898	0.903	0.868
	51	1.35	0.29	87	0.737	0.721	0.725
	52	1.35	0.59	52	0.441	0.397	0.427
	53	1.60	0.00	175	1.000	1.01	1.000
	54	1.60	0.15	154	0.880	0.874	0.861
	55	1.60	0.44	98	0.560	0.527	0.554
	56	1.60	0.59	70	0.400	0.392	0.394
	57	1.85	0.00	235	1.000	1.005	1.000
	58	1.85	0.29	165	0.702	0.698	0.695
	59	1.85	0.44	127	0.540	0.515	0.532
	60	1.85	0.59	89	0.379	0.386	0.365
	61	2.10	0.15	240	0.842	0.823	0.848
	62	2.10	0.29	194	0.681	0.686	0.683
	63	2.10	0.44	148	0.519	0.502	0.512
	64	2.10	0.59	103	0.361	0.381	0.338
Testing	65	0.350	0.000	276	1.000	1.000	1.000
	66	0.600	0.489	90	0.243	0.245	0.247
	67	0.850	0.367	210	0.382	0.387	0.380
	68	1.100	0.244	340	0.531	0.518	0.537
	69	1.35	0.12	450	0.900	0.946	0.892
	70	1.60	0.00	705	1.000	1.01	1.000
	71	1.85	0.489	370	0.451	0.461	0.478
	72	2.1	0.367	545	0.586	0.594	0.599
	73	0.35	0.294	44	0.518	0.511	0.513
	74	0.60	0.15	83	0.722	0.700	0.702
	75	0.85	0.00	145	1.000	0.999	1.000
	76	1.10	0.59	20	0.112	0.131	0.089
	77	1.35	0.44	69	0.585	0.540	0.578
	78	1.60	0.29	126	0.720	0.709	0.710
	79	1.85	0.15	202	0.860	0.846	0.854
	80	2.10	0.00	285	1.000	0.998	1.000

reinforcement layer ratio d_f/B. In the present study, out of 80 data points 64 points were used for training and remaining 16 were kept for testing. Each data point represents a complete laboratory model test on geogrid reinforced bed which was led to failure. All the inputs and output are normalized in the range of [−1, 1] before training.

A feed-forward back-propagation neural network is used with hyperbolic tangent sigmoid function and linear function as the transfer function. The network is trained with Levenberg-Marquardt (LM) algorithm as it is efficient in comparison to gradient descent back-propagation algorithm. The ANN has been implemented using MATLAB V 7.11.0(R2015b).

3 Results and Discussion

Two inputs and one output parameters were considered in the ANN model. The maximum, minimum, average and standard deviation values of the two input and one output parameters used in the ANN model are presented in Table 2.

Table 2. Statistical values of parameters

Parameter	Maximum value	Minimum value	Average value	Standard deviation
d_f/B	2.1	0.35	1.225	0.572
α/ϕ	0.588	0	0.269	0.192
RF	1	0.112	0.638	0.264

The schematic diagram of ANN architecture is shown in Fig. 2. The number of neurons in hidden layer is varied and it was selected based on the mean square error (MSE) value which was 0.001. In this ANN model four neurons are evaluated in hidden layer as shown in Fig. 3. Therefore the final ANN architecture is retained as 2-4-1 [i.e. 2 (input) – 4 (hidden layer neuron) – 1 (output)]. Mean square error (MSE) is defined as

$$MSE = \frac{\sum_{i=1}^{n} \left(RF_i - RF_p\right)^2}{n} \tag{2}$$

Coefficient of efficiency, R^2 is defined as

$$R^2 = \frac{E_1 - E_2}{E_1} \tag{3}$$

where,

$$E_1 = \sum_{i=1}^{n} (RF_i - \overline{RF})^2 \tag{4}$$

and

$$E_2 = \sum_{i=1}^{n} (RF_P - RF_I)^2 \tag{5}$$

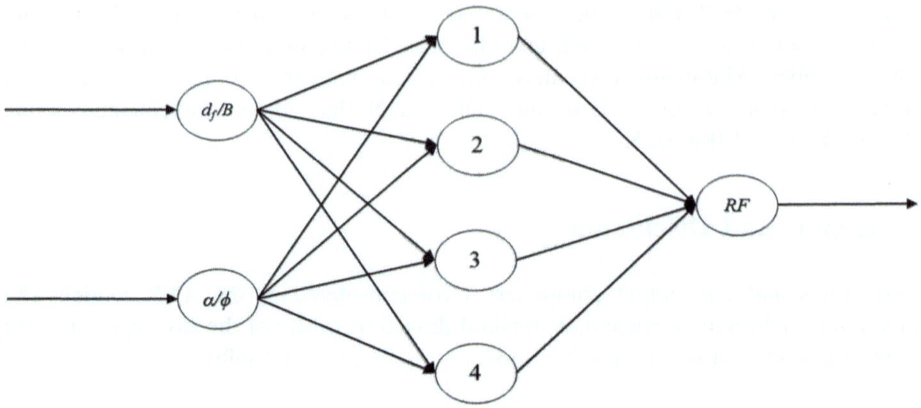

Fig. 2. Structure of ANN

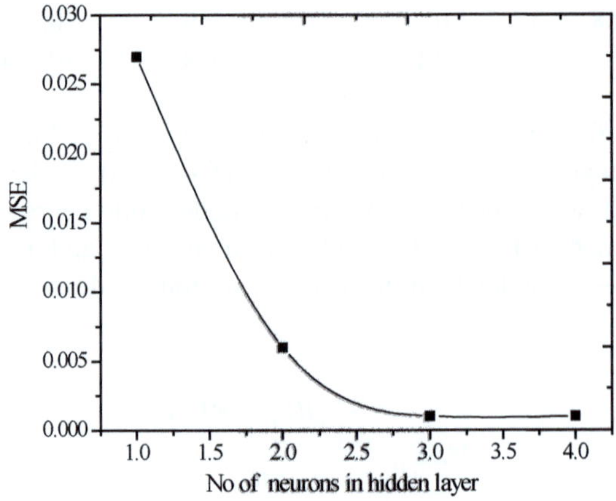

Fig. 3. Variation of hidden layer neuron with mean square error (mse)

where, RF_i, $\overline{RF}$ and RF_p are the experimental, average experimental, predicted RF values respectively; and n = number of training data.

The coefficient of efficiency (R^2) is found to be 0.9972 for training and 0.9952 for testing as shown in Figs. 4 and 5. All the data used in the training and testing have been obtained from laboratory model tests are from the same source and are of same nature. Probably, this may be one of the causes for better fitting in both training and testing phase as well. The weights and biases of the network are presented in Table 3. These weights and biases can be utilized for interpretation of relationship between the inputs and output, sensitivity analysis and framing an ANN model in the form of an equation. The residual analysis was carried out by calculating the residuals in between experimental reduction factor and predicted reduction factor for training data. Residuals can be defined as the difference between the experimental and predicted RF value and is given by

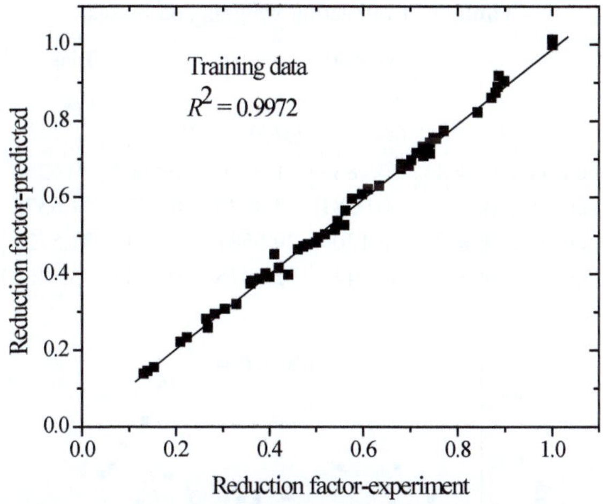

Fig. 4. Correlation between predicted reduction factor with experimental reduction factor for training data

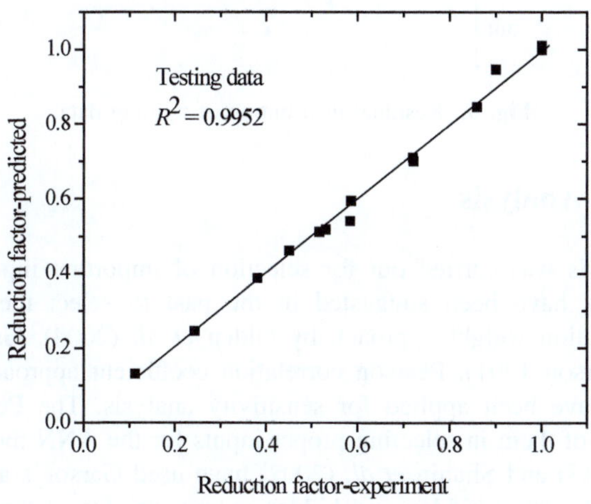

Fig. 5. Correlation between predicted reduction factor with experimental reduction factor for testing data

$$e_r = RF_i - RF_p \tag{6}$$

The residuals are plotted with the experimental number as shown in Fig. 6. It is observed that the residuals are evenly distributed along the horizontal axis of the plot. Therefore it can be said that the network is well trained and can be used for prediction with reasonable accuracy.

Table 3. Connection weights and biases

Neuron	Weight			Bias	
	w_{ik}		w_k	b_{hk}	b_O
	(d_f/B)	(α/ϕ)	RF		
Hidden neuron 1 (k = 1)	−0.1886	−1.5416	0.5663	0.6264	0.2388
Hidden neuron 2 (k = 2)	0.9381	3.8625	−0.2274	1.6575	
Hidden neuron 3 (k = 3)	13.1304	29.6581	−0.2938	31.5729	
Hidden neuron 4 (k = 4)	46.6441	−2.8076	0.3134	3.8462	

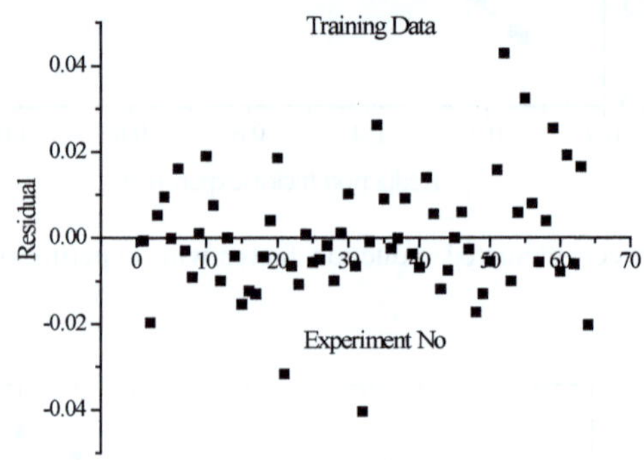

Fig. 6. Residual distribution of training data

4 Sensitivity Analysis

Sensitivity analysis was carried out for selection of important input variables. Different approaches have been suggested in the past to select the important input variables. Connection weight approach by Olden *et al.* (2004), Garson's algorithm approach by (Garson 1991), Pearson correlation coefficient approach by Guion and Elisseff (2003) have been applied for sensitivity analysis. The Pearson correlation coefficient is one of them in selecting proper inputs for the ANN model. Goh (1994), Behera et al. (2013) and Shahin *et al.* (2002) have used Garson's algorithm (Garson 1991) in which the input-hidden and hidden-output weights of trained ANN model are partitioned and the absolute values of weights are taken to select the important input variables. It does not provide information on the effect of input variables in terms of direct or inverse relation to the output. Olden *et al.* (2004) proposed a connection weights approach based on the NID, in which the actual values of input-hidden and hidden-output weights are taken. Table 4 shows the

Table 4. Cross-correlation of the input and output for the reduction factor

Parameters	d_f/B	α/ϕ	RF_{expt}
d_f/B	1	0	0.247
α/ϕ		1	−0.928
RF_{expt}			1

cross-correlation of inputs with the reduction factor (*RF*) value. It can be seen that *RF* is highly correlated to α/ϕ with a cross correlation value of 0.928, followed by d_f/B. From analysis of Garson's algorithm as presented in Table 5 it is seen that α/ϕ is found to be most Important input parameter with the relative importance value being 61.13% followed by 38.86% for d_f/B. Olden *et al.* (2004) proposed a connection weights approach based on the NID, in which the actual values of input-hidden and hidden-output weights are taken. It sums the products across all the hidden neurons, which is defined as S_i. The most important input corresponds to highest S_i value. As per connection weight approach analysis it is seen that α/ϕ is found to be most important input parameter (S_i value = −11.34) followed by d_f/B (S_i value = 10.44). The S_i values being negative imply that α/ϕ is indirectly and d_f/B is directly related to *RF* values. From the sensitivity analysis it can be seen that α/ϕ is found to be the most important parameter in predicting *RF*. In other words, increasing α/ϕ will lead to a reduction in the *RF* and hence leads to lower ultimate bearing capacity. Increasing d_f/B increases the *RF*, and hence, increases the bearing capacity.

5 Neural Interpretation Diagram (NID)

Ozesmi and Ozesmi (1999) proposed a neural interpretation diagram (NID) for visual interpretation of the connection weight among the neurons. For the present study with the weights as obtained and shown in Table 3, an NID is presented in Fig. 7. The lines joining the input-hidden and hidden-output neurons represent the weights. The positive weights are represented by solid lines and negative weights by dashed lines and the thickness of the lines is proportional to their magnitude. The input directly related to the output is represented with a grey circle and that having inverse effect with blank circle. It can be seen from Table 5 (4th column) that S_i value for parameter α/ϕ is negative indicating that the parameters α/ϕ is inversely related to *RF* values, whereas S_i value for parameter d_f/B being positive is directly related to *RF* values. The same has been shown in Fig. 7. Thus it is inferred that *RF* value decreases with increase in α/ϕ value and increases with increase in d_f/B value.

Table 5. Relative importance of different inputs as per Garson's algorithm and connection weight approach

Parameters	Garson's algorithm		Connection weight approach	
(1)	Relative importance (%) (2)	Ranking of inputs as per relative importance (3)	S_i values as per connection weight approach (4)	Ranking of inputs as per relative importance (5)
d_f/B	38.86	2	10.44	2
α/ϕ	61.13	1	−11.34	1

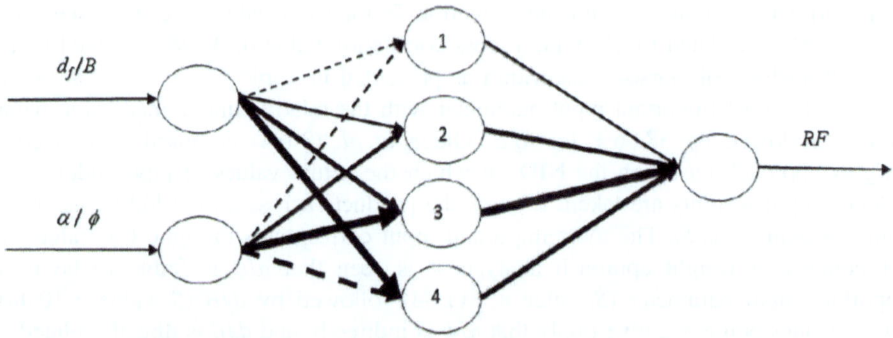

Fig. 7. Neural interpretation diagram showing lines representing connection weights and effects of inputs on reduction factor (RF)

6 ANN Model Equation for the Reduction Factor Based on Trained Neural Network

In the present study, with only two parameters (d_f/B and α/ϕ) a model equation is developed using the weights obtained from trained neural network model (Goh *et al.* 2005). The equation relating the input variables and the output can be written as,

$$RF_n = f_n\left\{b_0 + \sum_{k=1}^{h}\left[w_k f_n\left(b_{hk} + \sum_{i=1}^{m} w_{ik}X_i\right)\right]\right\} \tag{7}$$

where,
RF_n = normalized value of RF in the range [−1, 1]
f_n = transfer function
h = number of neurons in the hidden layer
X_i = normalized value of inputs in the range [−1, 1]
m = no. of input variables
w_{ik} = connection weight between i^{th} layer of input and k^{th} neuron of hidden layer
w_k = connection weight between k^{th} neuron of hidden layer and single output neuron
b_{hk} = bias at the kth neuron of hidden layer
b_o = bias at the output layer.

Using the values of trained weights and biases in Table 3, a step by step procedure is written down to form a relationship in the form of a equation between the input parameters (d_f/B and α/ϕ) and the output (RF).

- Step – 1
  The input parameters were normalized in the range [−1, 1] by the following expression

$$X_n = 2\left(\frac{X_1 - X_{\min}}{X_{\max} - X_{\min}}\right) - 1 \tag{8}$$

where,
X_n = normalized value of input parameter
X_{max} = maximum values of the input parameter
X_{min} = Minimum values of the input parameter
X_1 = is the data set.

- Step – 2
  Calculate the normalized value of reduction factor (RF_n) using the following expression

$$A_1 = -0.1886\left(\frac{d_f}{B}\right)_n - 1.5416\left(\frac{\alpha}{\varphi}\right)_n + 0.6264 \tag{9}$$

$$A_2 = 0.9381\left(\frac{d_f}{B}\right)_n + 3.8625\left(\frac{\alpha}{\varphi}\right)_n + 1.6575 \tag{10}$$

$$A_3 = 13.1304\left(\frac{d_f}{B}\right)_n + 29.6581\left(\frac{\alpha}{\varphi}\right)_n + 31.5729 \tag{11}$$

$$A_3 = 46.6441\left(\frac{d_f}{B}\right)_n - 2.8076\left(\frac{\alpha}{\varphi}\right)_n + 3.8462 \tag{12}$$

$$B_1 = 0.5663\left(\frac{e^{A_1} - e^{-A_1}}{e^{A_1} + e^{-A_1}}\right) \tag{13}$$

$$B_2 = -0.2274\left(\frac{e^{A_2} - e^{-A_2}}{e^{A_2} + e^{-A_2}}\right) \tag{14}$$

$$B_3 = -0.2938\left(\frac{e^{A_3} - e^{-A_3}}{e^{A_3} + e^{-A_3}}\right) \tag{15}$$

$$B_4 = -0.3134\left(\frac{e^{A_4} - e^{-A_4}}{e^{A_4} + e^{-A_4}}\right) \tag{16}$$

$$C_1 = -0.2683 + B_1 + B_2 + B_3 + B_4 \tag{17}$$

$$RF_n = C_1 \tag{18}$$

- Step – 3
  Denormalize the RF_n value obtained from Eq. 18 to actual RF as

$$RF = 0.5\,(RF_n + 1)\,(RF_{\max} - RF_{\min}) + RF_{\min} \tag{19}$$

$$RF = 0.5\,(RF_n + 1)\,(1 - 0.112) + 0.112 \tag{20}$$

where, RF_{max} = maximum value of *RF* in the database and RF_{min} = minimum value of *RF* in the database.

7 Comparison with Empirical Equation by Sahu *et al.* (2016)

Sahu *et al.* (2016) proposed an empirical equation based on laboratory model tests results for prediction of *RF*, which can be expressed as

$$\begin{aligned} RF &= \frac{q_{u(\alpha/\varphi, d_f/B)}}{q_{u(\alpha/\varphi=0,\ d_f/B)}} \\ &= \left[\left[1.36 - 0.45\left[\frac{D_f}{B}\right]\right]\left[\frac{d_f}{B}\right]^{\left[0.08\,+\,0.25\,\left[\frac{D_f}{B}\right]\right]}\left[\frac{\alpha}{\varphi}\right]^{\left[0.77\,+\,0.29\,\left[\frac{D_f}{B}\right]\right]}\right] \end{aligned} \tag{21}$$

where, $q_{uR(e/B,\, d_f/B)}$ = ultimate bearing capacity of geogrid reinforced sand due to inclined loading for a particular d_f/B; $q_{uR(\alpha/\varphi=0,\, d_f/B)}$ = ultimate bearing capacity of geogrid reinforced sand for $\alpha\,/\,\varphi = 0$ at the same d_f/B; and *RF* = Reduction factor.

As seen in Fig. 8 and Table 1, the comparison appears to be reasonably good. Hence, artificial neural network can be effectively used for the prediction of ultimate bearing capacity of strip footing in geogrid reinforced soil under inclined load.

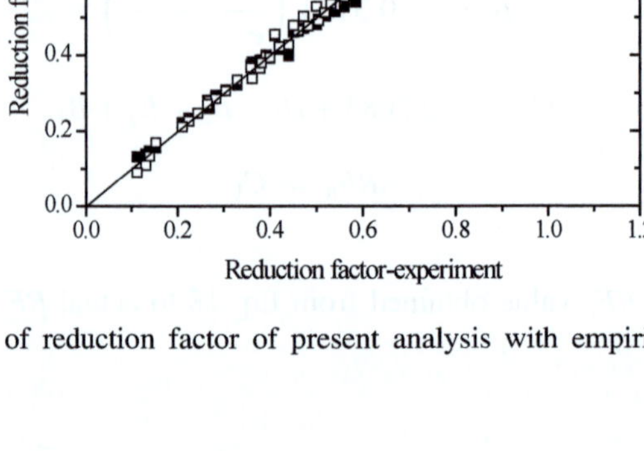

Fig. 8. Comparison of reduction factor of present analysis with empirical equation by Sahu *et al.* (2016)

8 Conclusions

The following conclusions can be drawn from the above study:

1. As per residual analysis, the errors are distributed evenly along the centerline. It can be concluded that the network is well trained and can predict the result with reasonable accuracy.
2. The developed ANN model could explain the physical effect of inputs on the output, as depicted in NID. It was observed that α/ϕ were inversely related to *RF* values, whereas, d_f/B was directly related to *RF*.
3. Based on sensitivity analyses; Pearson correlation coefficient, Garson's algorithm and connection weight approaches, it was observed that α/ϕ is the most important parameter.
4. An equation is presented based on the trained weights of the ANN.
5. The predictability of ANN models are found to be slightly better than the empirical equation developed by Sahu *et al.* (2016).

References

Behera, R.N., et al.: Prediction of ultimate bearing capacity of eccentrically inclined loaded strip footing by ANN part 1. Int. J. Geotech. Eng. (2013). doi:10.1179/1938636212Z.00000000012

Garson, G.D.: Interpreting neural-network connection weights. Artif. Intell. Expert **6**(7), 47–51 (1991)

Goh, A.T.C.: Seismic liquefaction potential assessed by neural network. J. Geotech. Eng. **120**(9), 1467–1480 (1994)

Goh, A.T., et al.: Bayesian neural network analysis of undrained side resistance of drilled shafts. J. Geotech. Geoenviron. Eng. **131**(1), 84 (2005). doi:10.1061/(ASCE)10900241

Guyon, I., Elisseeff, A.: An introduction to variable and feature selection. J. Mach. Learn. Res. **3**, 1157–1182 (2003)

Ozesmi, S.L., Ozesmi, U.: An artificial neural network approach to spatial modeling with inter specific interactions. Ecol. Model. (1999). doi:10.1016/S0304-3800(98)00149-5

Olden, J.D., et al.: An accurate comparison of methods for quantifying variable importance in artificial neural networks using simulated data. Ecol. Model. **178**(3), 389–397 (2004)

Shahin, M.A., et al.: Predicting settlement of shallow foundations using neural network. J. Geotech. Geoenviron. Eng. **128**(9), 785 (2002). doi:10.1061/(ASCE)1090-0241

Sahu, R., et al.: Bearing capacity of shallow strip foundation on geogrid-reinforced sand subjected to inclined load (2016). doi:10.1080/19386362.2015.1105622

Soft Soil Improvement with Conventional and Geogrid-Encased Stone Piles Under an Embankment

Mohamd B.D. Elsawy(✉)

Department of Civil Engineering, Faculty of Engineering,
Aswan University, Aswan, Egypt
melsawy@ut.edu.sa

Abstract. Construction on natural soft soil is considered a risk due to its low shear strength and permeability as well as its high compressibility. Stone piles technique have been utilized in soft soil to increase the bearing capacity and accelerates the consolidation. To improve the reinforcement and the drainage functions of the stone piles, the geosynthetic are used as encasement. In the current research, a case history of an embankment constructed on the reinforced soft soil with conventional stone piles has been chosen from the past research to be simulated. 3-Dimension, plane strain and axisymmetric techniques are used to simulate the embankment parts. The stone piles are reinforced by geogrid material to imply the influence of the encasement on the behavior of the stone piles-soft soil foundation. The consolidation analysis is applied to investigate the long-term behavior of the clay. There is a good agreement between the FEM results and the field measurements of the reinforced soft soil with conventional stone piles. The 3D and the axisymmetric models induces better agreement with field measurements than that of the plane strain model. The reinforced soft soil with encased stone pile has a smaller settlement and a shorter consolidation time than those of the reinforced soft soil with conventional stone piles. The reduction in the settlement is more significant with developing consolidation time. The dissipation of the excess pore water pressure in the reinforced clay with encased pile consumes shorter time in comparison with the reinforced clay with conventional piles. The effective vertical stress and the stress concentration in the encased piles are higher than those in the conventional piles. The encasement also causes reduction in the total stress of the surrounding clay which participates in the acceleration of the consolidation.

1 Introduction

Embankments constructed on soft deposits sustains problems such as base failure, excessive settlement, large lateral deformation, and local or global instability. Stone piles are increasingly used as an effective ground improvement method to support a wide variety of structures, including buildings and embankments (Balaam and Booker 1985; Lee and Pande 1998; Christoulas et al. 2000; Bae et al 2002; Borges et al. 2009; Murugesan and Rajagopal 2010). Stone piles accelerate consolidation of soft soil resulting from the generated lateral drainage paths, increased load-bearing capacity,

S.K. Shukla and E. Guler (eds.), *Advances in Reinforced Soil Structures*,
Sustainable Civil Infrastructures, DOI 10.1007/978-3-319-63570-5_10

and reduced settlement (Mitchell and Huber 1985; Han and Ye 1992, 2001; Bergado and Long 1994). The bearing capacity of stone piles depends mainly on the lateral support from the shear strength of the surrounding soil. If lateral support is not sufficient, the stone piles sustain excessive settlement and they may fails by lateral bulging. Therefore, construction of stone piles in soft soils with undrained strength (Su) less than 15 kPa (FHWA 1980) is almost impossible, however, because of the lack of lateral support.

Encasing stone piles with geosynthetics was recommended to enhance the lateral pile confinement which resulting on increasing the load-bearing capacity (Alexiew et al. 2005). Additionally, encasement prevents the lateral squeezing of stones into the surrounding soil and vice versa (Murugesan and Rajagopal 2006; Murugesan and Rajagopal 2007). The effectiveness of geosynthetic encasement on the capacity and settlement response of composite ground has been studied in laboratory and small-scale model tests (Sharma et al. 2004; Murugesan and Rajagopal 2007; Gniel and Bouazza 2010; Ali et al. 2012; Dash and Bora 2013). Successful numerical studies of encased stone piles can also be found in the literature. These analyses have historically been performed assuming either plane strain (e.g., Han et al. 2007; Tan et al. 2008; Zheng et al. 2009; Deb 2010) or axisymmetric (e.g., Tan et al. 2008; Borges et al. 2009; Almeida et al. 2013; Castro et al. 2013; Elsawy 2013; Ng and Tan 2014; Hosseinpour et al. 2015; Khabbazian et al. 2015; Ngoa and Tungb 2016) idealizations. The concept of a "unit cell", as axisymmetric idealization, has been the most popular approach for numerically simulating the response of either conventional or encased stone piles reinforced soft soil under embankments. While the 3D Modelling of embankments on conventional and encased stone pile reinforcing soft soil is very limited in the past research (Khabbazian et al. 2015). Additionally, Comparison between the 3D, unit cell and plain strain models wasn't done by the past researchers.

With regard to the practical applications of conventional and geosynthetic-encased stone piles under embankment with field measurements were done by (Tan et al. 2008; Almeida et al. 2014). Additionally, Raithel et al. (2002) also cited many successful case histories of geotextile-encased granular columns (GECs) under embankment in Germany, Sweden, and the Netherlands of the use of GECs for the stabilization of embankments on soft soils.

In the current research a case history of an embankment was chosen from the past research. The embankment constructed on reinforced soft soil with conventional stone piles has been numerically simulated with 3D, unit cell and plain strain models. The results of the field measurements and the numerical analyses were compared. The simulation of the embankment construction on reinforced soft soil with encased stone piles has also been done utilizing 3D and unit cell models. The consolidation behavior of this system is investigated to study the performance of the geogrid-encasement in the reinforced soil during and after consolidation. The results of the used three models have been compared considering the settlement, the excess pore water pressure, and the stresses in the pile and in the soil.

2 Case History Description

An embankment, located in Penchala Toll Plaza project at New Pantai Expressway, Malaysia, was adapted for the FEM simulation. A brief description of the project was given by Tan et al. (2008). The embankment geometry with the stone pile reinforced soft profile is shown in Fig. 1 having a line of symmetry on the left boundary. The 20 m wide and 1.8 m high embankment is filled by sandy material. The embankment has a slope of 2 horizontal to 1 vertical. The stone piles have a diameter of 0.8 m and a spacing distance between piles of 2.4 m. The stone piles, arranged in a square grid, extend through the soft soil for a depth of 6 m above the layer of the stiff clay. The upper crust layer is 1 m thick fill for strong soil. This layer was provided as a replacement of soft-clay surface to improve the ground for a stable construction platform as well as drainage of water during consolidation. The groundwater level is one meter below the ground surface. Two settlement plates (SP1 and SP2), as shown in Fig. 1, were installed in situ to measure the settlement at the center of the embankment and at 8 m from its toe. The excess pore water pressure was calculated at points A and B which located at a depth of 3.5 m. The points A and B have a horizontal distance of 1.2 m and 22 m far from the centerline of the embankment, respectively.

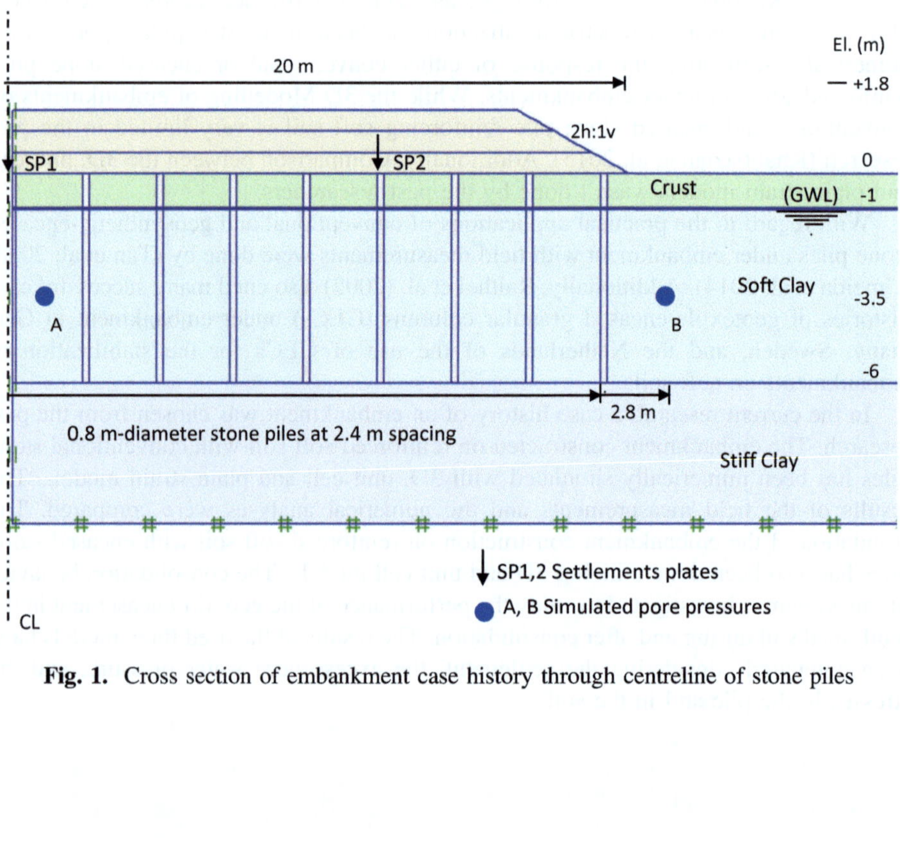

Fig. 1. Cross section of embankment case history through centreline of stone piles

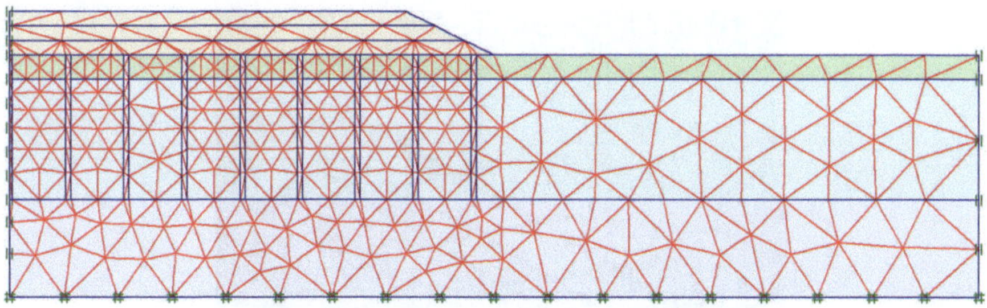

Fig. 2. Finite element mesh of the embankment case history through centreline of stone piles

3 Numerical Modeling and Selection of Parameters

2D and 3D modellings are utilized to simulate the embankment construction. The Mohr-coulomb model is used to describe all soils, soft soil, stone piles, crust layer, embankment fill and stiff clay, which is considered realistic approximations of the soils. The used parameters are depicted in Table 1. All the soils are modelled in drained conditions except the soft clay is modelled in undrained conditions. The project involved rapid embankment construction. The stone piles first installed by partial soil replacement. Then, the construction of the embankment was done three equal layers (0.6 m increment in embankment height in each layer). Each layer of the embankment is constructed with 3 days consolidation period, giving altogether 9 days. The consolidation process has been then continued with no change in loading condition until the remaining excess pore water pressure fell below a specified near zero value (1 kPa), which reach the end of the simulation.

Table 1. Material parameters for embankment models

Material	Unsatura-ted γ (kN/m³)	Saturated γ (kN/m³)	ν	E (kPa)	k_h (m/s)	k_v (m/s)	ć (kPa)	ϕ (deg)
Embankment Fill	18	20	0.3	15,000	1.16×10^{-5}	1.16×10^{-5}	3	33
Crust	17	18	0.3	15,000	3.47×10^{-7}	1.16×10^{-7}	3	28
Soft clay	15	15	0.3	1,100	3.47×10^{-9}	1.16×10^{-9}	1	20
Stiff clay	18	20	0.3	40,000	3.47×10^{-9}	1.16×10^{-9}	3	30
Stone pile	19	20	0.3	30,000	1.16×10^{-4}	1.16×10^{-4}	5	40

CSP: Conventional Stone Piles
ESP: Encased Stone Piles

The stone piles are assumed to be encased with Combigrid 40/40 Q1 (Naue GmbH). The composite geogrid/non-woven geotextile is a geogrid covered by a geotextile to allow drainage without mixing soft soil with stone particles, as illustrated in Fig. 3. The geotextile is arranged in such a way that it would not contribute either to the vertical or radial stiffness of the encased stone pile. The geogrid encasement is modeled as a linear elastic continuum element with a series of one-dimensional bare

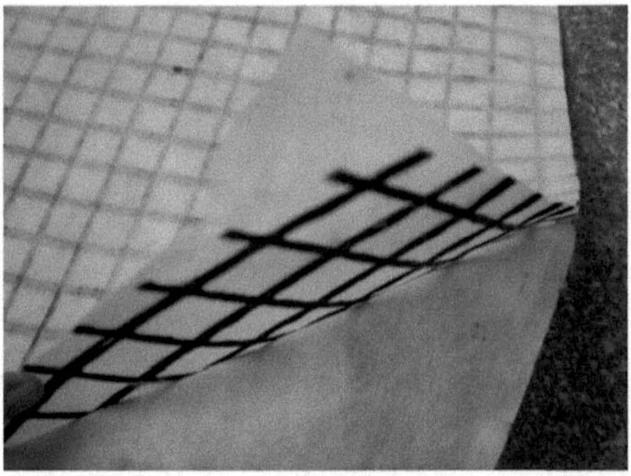

Fig. 3. Composite of geogrid/non-woven geotextile

(line) elements having no bending stiffness, using flexible elastic elements which can mobilize only axial tension forces. The elastic parameter used in modelling the geogrid encasement with Plaxis program is only the axial stiffness J = EA (forces per unit width per unit strain). The encasement is subjected to axial extension, and there is no other deformations: therefore the Poisson's ratio of the encasement equals zero. The geogrid stiffness (J = EA) is calculated at a strain of 2% where geogrid is under working stress conditions. The used geogrid has a stiffness of J = 800 kN/m. No interaction between the geogrid and the surrounding soil was assumed in the current study. This is because no slippage occurs between the geogrid and the surrounding soil, Elsawy (2013).

3.1 2D Modelling

The plane-strain modelling is possible as the embankment extended to a distance of more than 200 m with approximately uniform cross-sectional geometry. Hence, in the current study the plane-strain modelling has been done by Plaxis 2D Version 8.1 using width of the stone piles (or walls) of 0.21 m and spacing distance of 2.4 m, as illustrated in Figs. 1 and 2. The plane-strain pile width is given by the following relationship based on the equivalence area of the replacement ratio:

$$b_c = B\left(r_c^2/r_e^2\right) \tag{1}$$

Where b_c is the half width of the wall, B = (S/2) is the half of the spacing distance between piles, r_c = (d/2) is the radius of the stone pile and r_e = (d_e/2) is the radius of the drainage zone or the unit cell which is equivalent to the plane strain width. Where is in square orientation r_e = 1.13 B, as shown in Fig. 4a.

The unit cell technique has been also used in this study to simulate the embankment using axisymmetric model in Plaxis 2D. The unit cell consists of a stone pile and the surrounding soft soil over the stiff clay, as shown in Fig. 4. The stone piles are installed in square panels which produces an equivalent unit cell with a diameter of 2.72 m

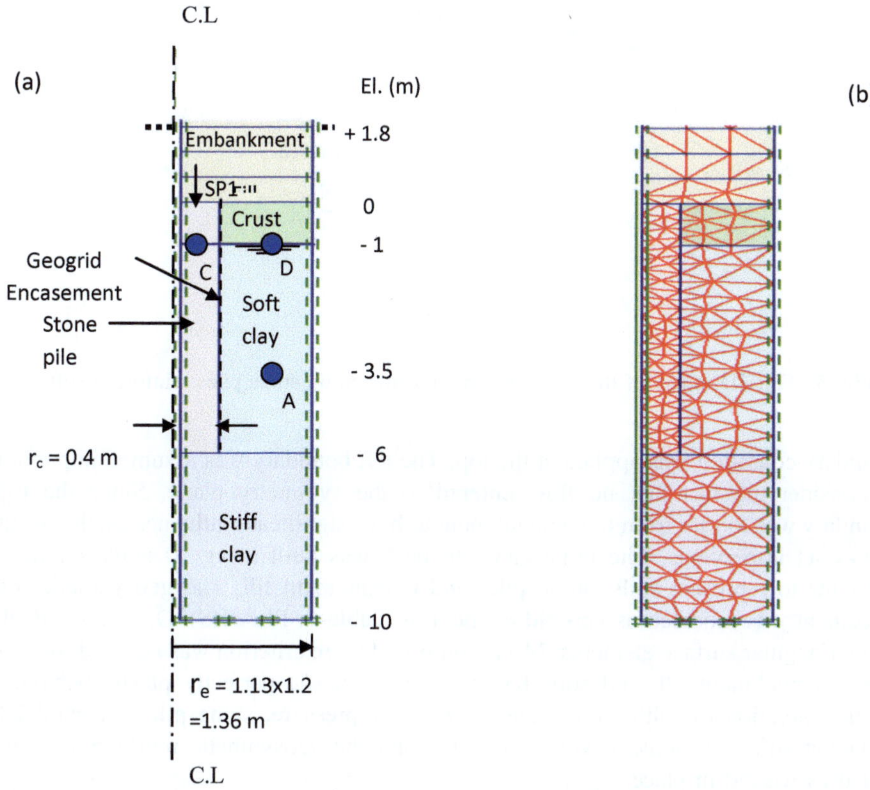

Fig. 4. Unit cell (a) Model Parts (b) Finite element mesh

(1.13 × 2.4 m). In 2D modelling (the plain strain and the axisymmetric models), half of the model is simulated utilizing medium mesh with 15-node wedge elements as illustrated in Figs. 2 and 4b. The horizontal and the vertical displacements are restrained in the bottom boundaries while the horizontal displacement is only restrained in the lateral boundaries.

3.2 3D Modelling

Full 3D model was developed to understand the long-term behaviour of stone piles and geogrid-encased stone piles-reinforced ground. The commercial FE package PLAXIS 3D (2012) was used for the modelling. Half of the model is simulated as shown in Fig. 5. Relatively medium mesh arrangements were used. The mesh was refined in the region of the pile soil interface to increase the accuracy of the predictions.

As displacement boundary is concerns, no displacements in the directions perpendicular to the symmetry planes and to the base were allowed. For the hydraulic boundary condition, the phreatic level was set at the top surface of the soft clay layer to generate a hydrostatic pore water pressure profile in the domain. A zero pore pressure

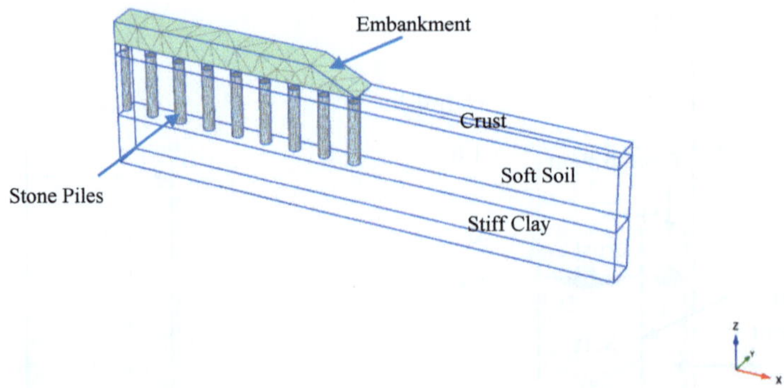

Fig. 5. Full 3D model of the embankment constructed on stone piles reinforced soft soil

boundary condition was applied at the top. The left boundary was assumed impervious to consider the fact that no flow entered to the symmetry plane. Since the right boundary was too far from the embankment to have significant influence on the results, it was set impervious. The finite element mesh was built using 10 node tetrahedral elements to represent soils, stone pile, and embankment fill. The geosynthetic reinforcement was modeled as geogrid element available in PLAXIS 3D, composed of 6 node triangular surface elements. Mohr-coulomb failure criterion was adopted for stone piles, embankment fill, and soils having linearly elastic perfectly plastic behaviour. After generation of initial stress and pore water pressure, stone pile was model by replacing soft soil element with stone pile, and the geosynthetic reinforcement was added as wished in place.

4 Discussion of the Results

The results of the numerical analyses having different modeling are compared to the field measurements of the settlement at SP1 and SP2. Additionally, the results of the excess pore water pressure from the current numerical analyses are compared to the results from 3D modeling of Tan et al. (2008) at point A and B. Tan et al. (2008) also simulated the embankment parts by the Plaxis 3D Tunnel Version 2 using 15-node wedge elements. Owing to the software limitations, the stone piles in three dimensional model were given as equivalent geometry with square cross-sectional area in place of the actual circular geometry.

4.1 Settlement

The settlement was calculated at SP1 in the 3D, in the plane strain and in the unit cell models to compare their results with the field measurements. Figure 6 shows the relationship of the settlement with time for the field data and the models of the embankment at SP1. The settlement increases with time with rapid rate until the time reaches approximately 35 days. After that time, the settlement increases along time with a very small rate. The settlement hasn't approximately any increase after the time of 90 days. This means that the consolidation finished.

There is a good agreement between the FEM results and the field measurements. The plane strain model induces settlement more than the field measurement. In the other hand, the 3D model and the unit cell model for the conventional stone piles imply settlement very close to the field settlement. Hence, the 3D and the axisymmetric model induce better agreement with field measurements than that of the plane strain model.

The stone piles are encased with geogrid material which has a Stiffness of J = 800 kN/m. The embankment over the reinforced soil with encased stone piles is also modelled by the 3D and the unit cell models. The reinforced soft soil with encased stone pile has a settlement smaller than that of the reinforced soft soil with conventional stone pile. Figure 6 shows also that the consolidation settlement is accelerated when encasing the stone piles. The 3D and the unit cell models induce a very good agreement for the settlement not only in the reinforced soil with conventional piles but also in the reinforced ground with encased piles.

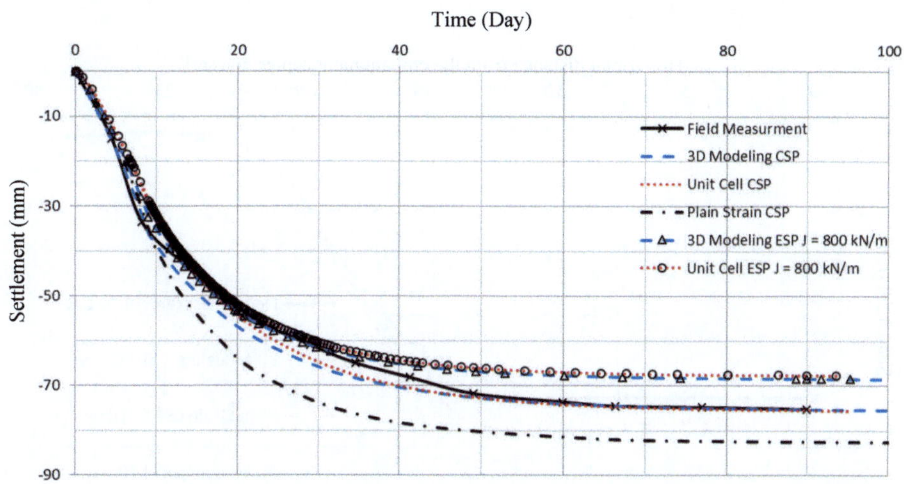

Fig. 6. Time – settlement relationships for field measurements and numerical analyses

Figures 7 and 8 show the settlement along the horizontal distance from the embankment centreline after 20 days and 90 days since the embankment has been constructed, respectively. The field measurements are compared with the FEM models. The measured settlement at SP1 and SP2 is approximately the same which leads to that the settlement under the embankment is constant. The constant settlement is well agreed by the 3D modelling after both times of 20 days and 90 days. While the plane strain model induces an overestimation for the field settlement after 20 and 90 days. The 3D and the plane strain model also predicts the settlement distribution under the entire embankment. Beyond the embankment toe, the settlement converts to heave. The heave decreases with increasing time in which the two models are well agreed.

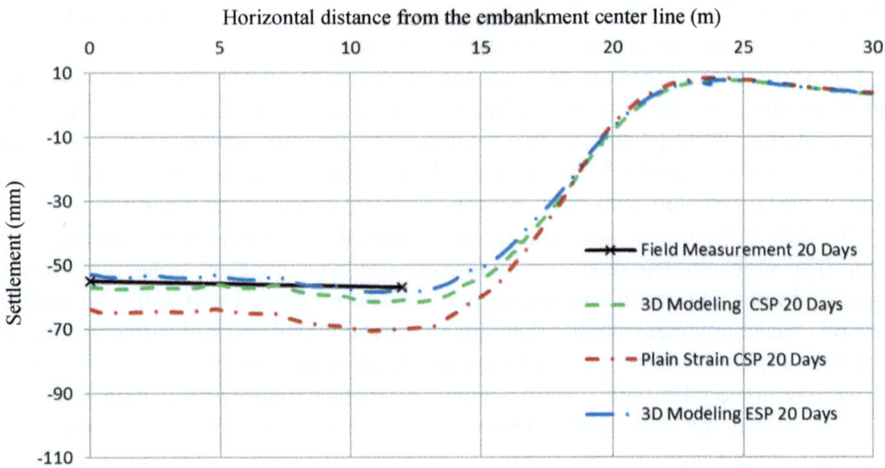

Fig. 7. Surface settlement at time 20 days after construction

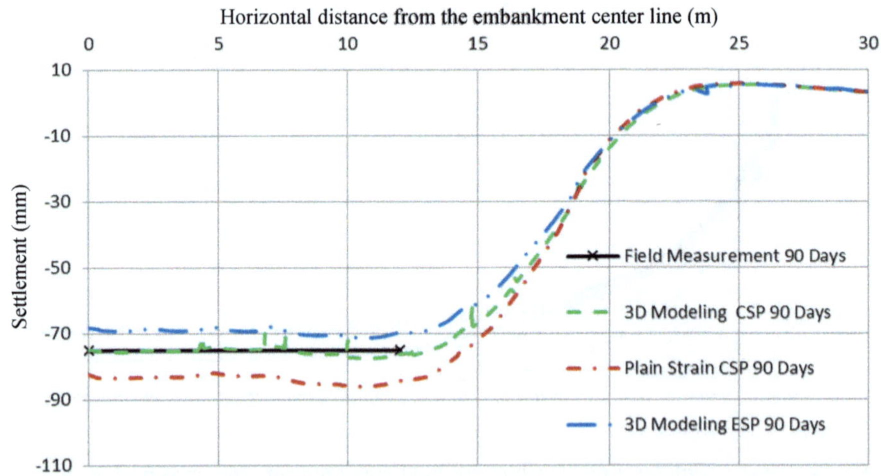

Fig. 8. Surface settlement at time 90 days after construction

When the stone piles are encased by geogrid material, a reduction only in the settlement occurs after 90 day from the construction. In the other hand, there is no significant reduction in the settlement after 20 days from construction. Hence, the encasement has a greater influence on the settlement reduction after the consolidation in comparison with the period directly after the construction. This phenomenon is due to the encasement increases the pile stiffness resulting in increase the stress transfer and stress concentration in the piles during the consolidation. In the other hand, the reinforced soft soil with encased piles causes a slight reduction in the settlement. The reasons of that are the applied load and the encasement stiffness are somewhat small.

4.2 Excess Pore Water Pressure

The excess pore water pressure was calculated from the 3D and the Plane strain model at locations A and B compared with the 3D simulation from the study of Tan et al. (2008). In the axisymmetric model, the excess pore water pressure was only calculated from at point A. Figures 9 and 10 show the simulated excess pore water pressure with time at locations A and B, respectively. The excess pore water pressure at A in the models has an initial peak value of approximately 17 kPa due to the embankment construction and then dissipates with different rates to nearly zero after 90 days. The 3D model, the plane strain model and the unit cell model induce somewhat lower values of the excess pore water pressure than those of Tan et al. (2008). In general, the models have good agreement between them. In the other hand, when the piles are encased with geogrid material, the dissipation of the excess pore pressure is slightly accelerated. This is more pronounced in the unit cell model as depicted in Fig. 9. The slight acceleration is because of the applied loads are small. It is expected that, the improvement in the production and in the dissipation of the excess pore presses will occur with increasing applied loads.

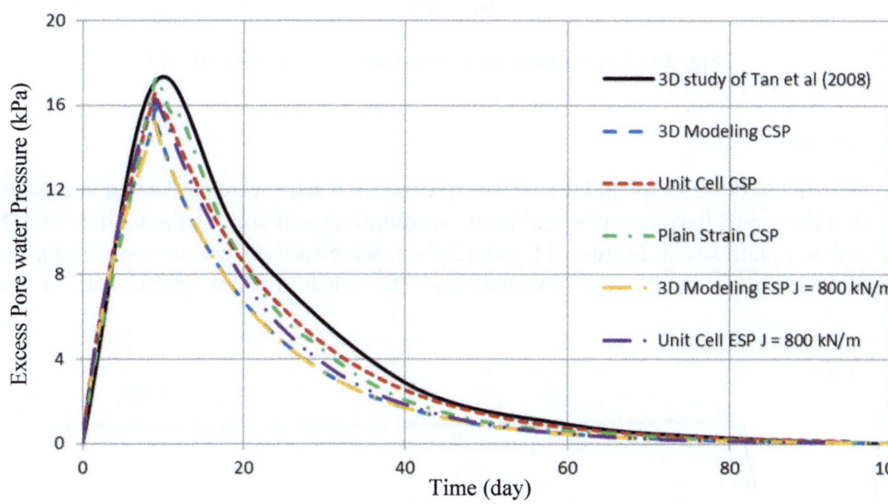

Fig. 9. Excess pore water pressure values at point A

On the other side, at point B, 2 m away from the embankment edge, the excess pore water pressure was calculated by the 3D and plane strain models. The two models show a good agreement for the distribution of the excess pore water pressure with the results of 3D simulation of Tan et al. 2008. The excess pore water pressure has significantly lower peak values due to the diminished effects of the embankment loading and dissipates almost identically in the models. This means that the excess pore water pressure discrepancies are only confined within a distance of several meters from the stone piles. The excess pore water pressure here takes much longer than 120 days to dissipate and hence the acceleration of consolidation by the stone piles is hardly evident at this location. Additionally, the pile encasement has a very weak effect on the excess pore water pressure of the soil as illustrated in Fig. 10.

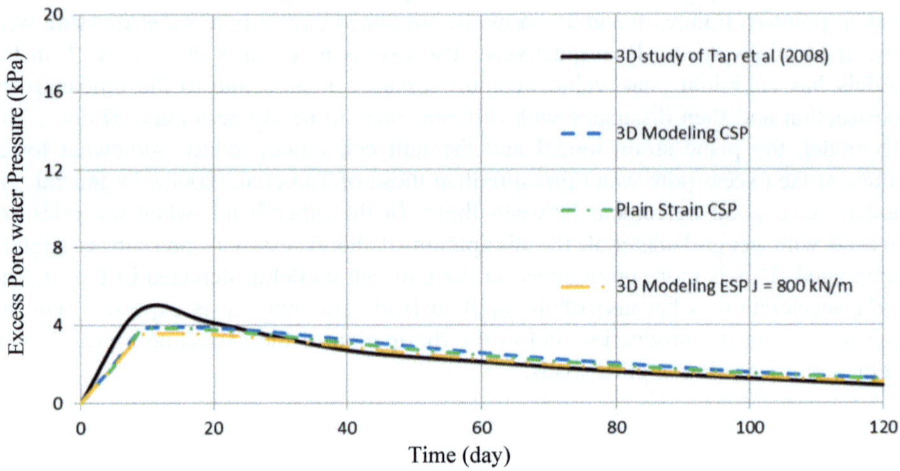

Fig. 10. Excess pore water pressure values at point B

4.3 Stress

The vertical effective stress and the stress concentration ratio were calculated at a depth of 1.0 in the top of the stone pile and in the surrounding soft soil for the models of 3D, unit cell and plain strain. Figures 11 and 12 show the effective vertical stress along the consolidation time in the conventional and the encased stone piles, and in the

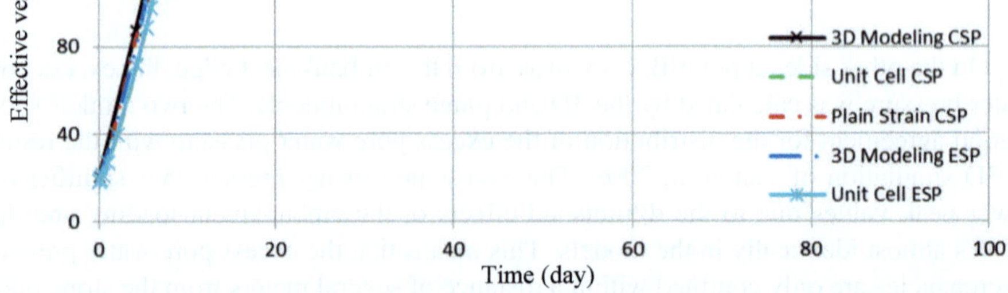

Fig. 11. Effective vertical stress in the top of the pile close to the embankment centre

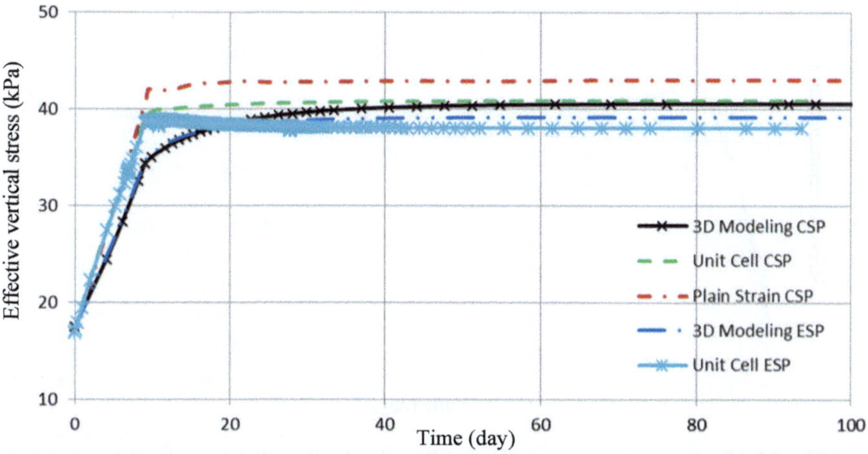

Fig. 12. Effective vertical stress in the top of the soil near the embankment centre

surrounding soil, respectively. The trends of the effective stress of the conventional and the encased stone piles in the clay are approximately similar. The developments of the effective stress with time in the reinforced clays are also approximately similar. The unit cell and plain strain models imply close results of the effective vertical stress and stress concentration ratio in both piles and soil. While the 3D model induces distant values in comparison with the other models.

The effective vertical stress in the encased stone piles is higher than that in the conventional stone piles. In contrast, the encasement of the stone pile causes reduction in the effective stress in the surrounding soft soil. This is pronounced in the 3D and the unit cell models. Additionally, beyond the yield point in the effective stress of the pile, the conventional piles induce somewhat softening while the encased piles imply somewhat hardening. Vice versa occurred in the surrounding soil, beyond the yield point of the effective stress, using conventional piles induces somewhat hardening while utilizing encased piles implies somewhat softening. These results agree with the study results of Almeida et al. (2014). This phenomenon is due to the stress transfer from the surrounding soft soil to the encased stone pile. The increase in the stiffness of the overall encased stone pile leads to increase the stress concentration in the pile and to increase also the stress transfer from the surrounding soil, as depicted in Fig. 13. The development of the stress concentration ratio is similar to that of the effective stress. 3D model implies greater values of stress concentration ratio for the reinforced soil with both the conventional and encased pile compared with 2D models. These results agree with the study of Ng and Tan (2014). The stress concentration values are ranged from 3 to 5 which agreed with values from field measurement and past studies (range from 2 to 9). The stress concentration phenomenon has an important role in reducing consolidation settlement, accelerating consolidation time and increasing bearing capacity of the reinforced soil.

The total stress in the reinforced soft soil with the conventional and the encased stone piles was calculated at point A which is located at a depth of 3.5 m. The

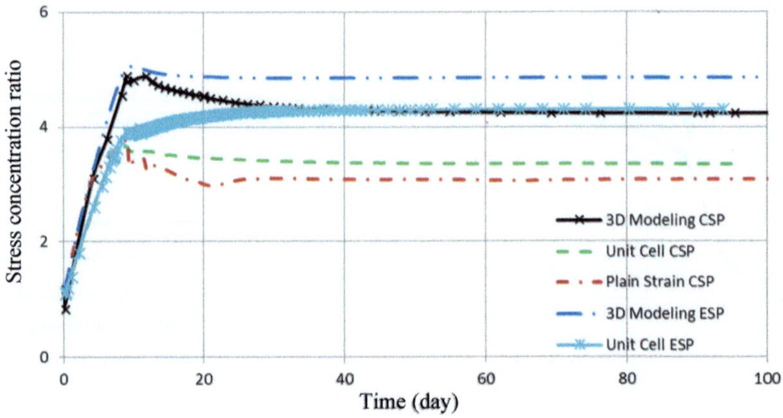

Fig. 13. Stress concentration ratio at 1.0 m depth close the embankment centre

relationship of the total vertical stress with consolidation time is shown in Fig. 14. The total vertical stress in the reinforced soil increases during construction until reaching maximum value. Beyond the maximum value, the reinforced soil with conventional piles implies approximate constant values of total stress along consolidation which is pronounced in the plain strain model. While the total stress decreases along consolidation when reinforcing soft soil with encased piles. The values of the total stress along consolidation in the reinforced soil with encased piles are smaller than those of the reinforced soil with convectional piles. This is more cleared in the unit cell model. This phenomenon means that greater load at the beginning and then the load decreases along consolidation time. The reinforced soil induces some degree of preloading. This phenomenon is not found in the unreinforced soft soil which normally has constant total

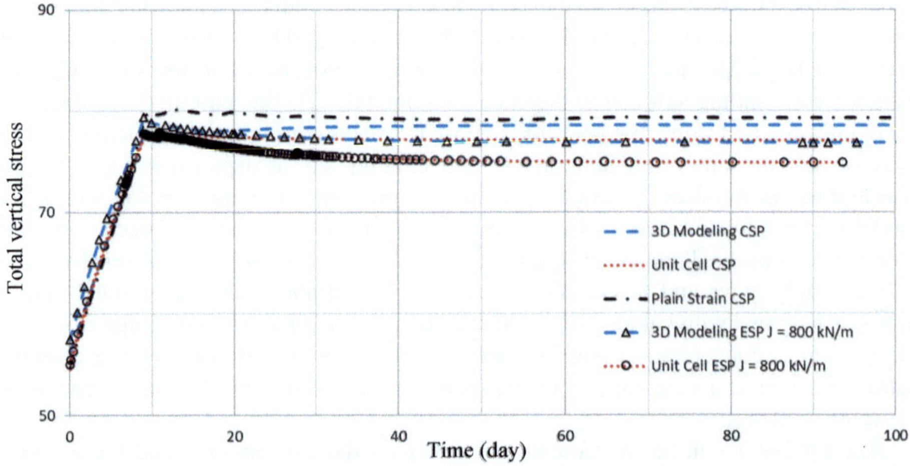

Fig. 14. Total vertical stress in soil at point A

stress during consolidation. The reduction in the total stress of the reinforced soft soil with encased stone pile is a result for the stress transfer from the soil and stress concentration in the piles (Elsawy 2013).

5 Conclusions

The construction of an embankment with 1.8 m height on reinforced soft soil with conventional stone piles is numerically simulated in the current study using the 3D, unit cell and plain strain models. The results of the numerical analyses are compared with the field measurement. The 3D and the unit cell models shows a vey good agreement with field measurement of the settlement while the plain strain model show overestimated results. The numerical analyses also investigated the behaviour of the reinforced soft soil with encased stone piles under the embankment. The stone piles are encased with geogrid material. The numerical results of the reinforced soil with conventional and encased piles show a very good agreement for the settlement between the 3D models and the unit cell model. The results of the excess pore water pressure imply a good agreement between all the used models. While the used models induce distant values of the stresses. This is more pronounced between the 3D model and the unit cell model. In the other side, the used models show the same developments of the settlement, the excess pore water pressure and the stresses along consolidation time. Hence, the results of the numerical analyses indicate that:

- Most of the consolidation settlement occurred at time of 35 days from construction. The reinforced soft soil with encased piles implies smaller settlement and consolidation time in comparison with the reinforced soft soil with conventional piles. The performance of the encased stone piles become better as developing consolidation time.
- There is a small acceleration in the dissipation of the excess pore water pressure when reinforcing soft soil with encased piles. This is attributed to that the level of the applied load is somewhat low.
- As the piles are encased, the stress concentration ratio increases. Additionally, the encasement of the piles doesn't only convert the softening in the effective stress of the pile to hardening but also do the vice versa in the effective stress of the surrounding soil. The pile encasement causes also reducing the total stress in the soil along consolidation which results in acceleration the consolidation time.

References

Alexiew, D., Brokemper, D., Lothspeich, S.: Geotextile encased columns (GEC): load capacity, geotextile selection and pre-design graphs. In: Proceedings of Geo-Frontiers Conference on ASCE, Reston, VA, pp. 497–510 (2005)

Ali, K., Shahu, J.T., Sharma, K.G.: Model tests on geosynthetic-reinforced stone columns: a comparative study. Geosynth. Int. **19**(4), 292–305 (2012)

Almeida, M.S.S., Hosseinpour, I., Riccio, M.: Performance of a geosynthetic-encased column (GEC) in soft ground: numerical and analytical studies. Geosynth. Int. **20**(4), 252–262 (2013)

Almeida, M.S.S., Hosseinpour, I., Riccio, M., Alexiew, D.: Behavior of geotextile-encased granular columns supporting test embankment on soft deposit. J. Geotech. Geoenviron. Eng. ASCE (2014). ISSN 1090-0241/04014116

Bae, W., Shin, B., An, B.: Behavior of foundation system improved with stone columns. In: Proceedings of the 12th International Offshore and Polar Engineering Conference, the International Society of Offshore and Polar Engineering, Kitakyushu, Japan, pp. 675–678 (2002)

Balaam, N.P., Booker, I.R.: Effect of stone column yield on settlement of rigid foundations in stabilized Clay. Int. J. Numer. Anal. Methods Ceomechanics **9**, 331–351 (1985)

Bergado, D.T., Long, P.V.: Numerical analysis of embankment on subsiding ground improved by vertical drains and granular piles. In: Proceeding of the XIII ICSMFE, New Delhi, India, pp. 1361–1366 (1994)

Borges, J.L., Domingues, T.S., Cardoso, A.S.: Embankments on soft soil reinforced with stone columns: numerical analysis and proposal of a new design method. J. Geotech. Geol. Eng. **27** (6), 667–679 (2009)

Castro, J., Sagaseta, C., Cañizal, J., Da Costa, A., Miranda, M.: Foundations of embankments using encased stone columns. In: Proceedings of the 18th International Conference on Soil Mechanics and Geotechnical Engineering, Paris, France (2013)

Christoulas, S.T., Bouckovalas, G., Giannaros, C.H.: An experimental study on model stone columns. Soils and Foundations Journal **40**(6), 11–22 (2000)

Dash, S.K., Bora, M.C.: Influence of geosynthetic encasement on the performance of stone columns floating in soft clay. Can. Geotech. J. **50**(7), 754–765 (2013)

Deb, K.: A mathematical model to study the soil arching effect in stone column-supported embankment resting on soft foundation soil. Appl. Math. Model. **34**(2010), 3871–3883 (2010)

Elsawy, M.B.: Behaviour of soft ground improved by conventional and geogrid-encased granular piles based on FEM study. Geosynth. Int. **20**(4), 276–285 (2013)

FHWA (Federal Highway Administration) "Highway sub-drainage design manual." Report TS-80-224, Washington, DC (1980)

Gniel, J., Bouazza, A.: Construction of geogrid encased stone columns: a new proposal based on laboratory testing. Geotext. Geomembr. **28**(1), 108–118 (2010)

Han, J., Ye, S.L.: Settlement analysis of buildings on the soft clays stabilized by stone columns. In: Proceedings of the International Conference on Soil Improvement and Pile Foundations, Nanjing, China, pp. 446–451 (1992)

Han, J., Ye, S.L.: Simplified method for consolidation rate of stone column reinforced foundations. ASCE J. Geotech. Geoenviron. Eng. **127**(7), 597–603 (2001)

Han, J., Oztoprak, S., Parsons, R.L., Huang, J.: Numerical analysis of foundation columns to support widening of embankments. Comput. Geotech. **34**(6), 435–448 (2007)

Hosseinpour, I., Almeida, M.S.S., Riccio, M.: Full-scale load test and finite-element analysis of soft ground improved by geotextile-encased granular columns. Geosynthetics Int. (2015). [http://dx.doi.org/10.1680/gein.15.00023]

Khabbazian, M., Kaliakin, V.N., Meehan, C.L.: Column supported embankments with geosynthetic encased columns: validity of the unit cell concept. Geotech. Geol. Eng. **33**, 425–442 (2015)

Lee, J.S., Pande, G.N.: Analysis of stone-column reinforced foundations. Int. J. Numer. Anal. Methods Geomech. **22**(12), 1001–1020 (1998)

Mitchell, J.K., Huber, T.R.: Performance of a stone column foundation. J. Geotech. Eng. **111**(2), 205–223 (1985)

Murugesan, S., Rajagopal, K.: Geosynthetic-encased stone column: numerical evaluation. Geotext. Geomembr. **24**(6), 349–358 (2006)

Murugesan, S., Rajagopal, K.: Model tests on geosynthetic encased stone columns. Geosynth. Int. **14**(6), 346–354 (2007)

Murugesan, S., Rajagopal, K.: Studies on the behavior of single and group of geosynthetic encased stone columns. J. Geotech. Geoenviron. Eng., 129–139 (2010). 10.1061/(ASCE)GT.1943-5606.0000187

Ng, K.S., Tan, S.A.: Stress transfer mechanism in 2D and 3D unit cell models for stone column improved ground. Int. J. Geosynth. Ground Eng. **1**, 3 (2014). doi:10.1007/s40891-014-0003-1

Ngo, N.T., Tung, T.M.: Coupled discrete-continuum method for studying load-deformation of a stone column reinforces rail track embankments. Procedia Eng. **142**, 138–144 (2016)

Raithel, M., Kempfert, H.G., Kirchner, A.: Geotextile encased columns (GEC) for foundation of a dike on very soft soils. In: Proceedings of 7th International Conference on Geosynthetics, Balkema, Nice, France, 1025–1028 (2002)

Sharma, R.S., Kumar, B.P., Nagendra, G.: Compressive load response of granular piles reinforced with geogrids. Can. Geotech. J. **41**(1), 187–192 (2004)

Tan, S.A., Tjahyono, S., Oo, K.K.: Simplified plane-strain modeling of stone-column reinforced ground. J. Geotech. Geoenviron. Eng. **134**(2), 185–194 (2008)

Zheng, J.J., Chen, B.G., Lu, Y.E., Abusharar, S.W., Yin, J.H.: The performance of an embankment on soft ground reinforced with geosynthetics and pile walls. Geosynth. Int. **16** (3), 173–182 (2009)

Analysis and Design of Piled Geogrid-Reinforced-Earth Embankment

Nasr O. Sheta(✉) and Rudolph P. Frizzi

Langan Engineering and Environmental Services, Inc., Parsippany, NJ, USA
{Nsheta, rfrizzi}@langan.com

Abstract. A pile-supported geogrid-reinforced-earth embankment was designed and constructed to serve as an access road for heavily loaded cranes and pile-installation rigs for constructing a major bridge between New York and New Jersey in the United States. The road is to also serve as an access to maintain the bridge during its service life. Subsurface investigations performed along the proposed access road alignment revealed subsurface conditions typically consisting of surficial fill underlain by a highly compressible organic clay and/or peat overlying marine sand underlain by glacial till. A limiting-settlement criterion set forth by the crane engineer together with settlement and slope-stability analyses indicated a conventional embankment could experience non-tolerable settlement and slope instability where thick organic soils were encountered. A ground-improvement program was carried out using geogrid reinforcement, a load transfer platform (LTP), and timber elements to control settlement and enhance slope stability. Where thinner organic soils were encountered, staged construction and a monitoring program were implemented to design and construct the proposed embankment.

1 Introduction

Design and construction of embankments on deep soft soils present challenges to geotechnical engineers because of potential long-term settlement and long construction time required to improve the soft soils as well as the relatively high cost associated with soft soil treatment/improvement. Pile-supported embankments provide a practical and efficient solution for embankment construction on soft soils because of shorter construction time. However, in some cases with significantly deep soft deposits, the potential for long-term settlement can be a considerable design concern, even when pile support is considered. The use of geosynthetics-reinforced LTPs above piles enhances load transfer from the embankment to piles, minimizes loads transmitted to soft soils and, hence, reduces total and differential settlements at the base of the embankment (Anjana and Rajagopal 2012). Geosynthetics-reinforced pile-supported embankment construction also results in significantly shorter construction time, compared to time needed to perform ground improvement of soft soils, limits lateral soil pressures on adjacent structures and significantly reduces, or even eliminates, embankment settlement (van Eekelen et al. 2015).

The design of a geosynthetics-reinforced pile-supported embankment includes design of embankment geometry, which is determined based on construction requirements, integrity of the embankment soil mass, which is typically evaluated

S.K. Shukla and E. Guler (eds.), *Advances in Reinforced Soil Structures*,
Sustainable Civil Infrastructures, DOI 10.1007/978-3-319-63570-5_11

based on slope stability analysis, and load transfer from embankment to the underlying pile-soil system, which is mainly based on soil arching, and load transfer through geosynthetics (Syawal et al. 2007). Load transfer from embankment to the underlying pile-soil system was explored both analytically and experimentally by several investigators. An approach presented by Carlsson (1987) considered a 2-D model with a triangular soil wedge centered between pile rows and transmitting its weight directly to geosynthetics and the underlying soil subgrade, while the weight of the rest of embankment soil and surcharge loads on top of embankment are directly transmitted to piles. Another approach was presented by Christopher (2014) to study the load transfer mechanism of geosynthetics-reinforced pile-supported embankments considering soil settlement and strain compatibility of the LTP. Svanø et al. (2000) developed a 3-D model to calculate the load sharing between piles and geosynthetics-subgrade system. Full-scale field tests and field instrumentation of test embankments were also performed to study behavior of geo-synthetics-reinforced pile-supported embankments (Laurent et al. 2008; Yan and Xiaoyan 2015). In addition, Poulos (2007) developed design charts of piles supporting embankments on soft clay. These design charts addressed different aspects of pile design including pile ultimate capacity, settlement and the effect of pile position below embankment. Navin (2005) also utilized finite element method (FEM) to study the stability of embankments founded on soft soils improved with deep-mixing columns. Such study recommended the use of FEM rather than limit equilibrium methods (LIMs), as embankment failure mechanisms resulting from column bending and tilting can take place, which can't be predicted by LIM.

This paper presents the analysis, design, construction and monitoring of a geosynthetics-reinforced-earth pile-supported embankment serving as an access road for heavily loaded cranes and pile-installation equipment utilized to construct a major bridge between New York and New Jersey in the United States.

2 Project Description and Subsurface Conditions

The subject site, where a \$1.5B (USD) cable-stayed double-deck bridge is to be constructed to replace an existing bridge, is located between New York and New Jersey in the United States. The subject pile-supported geogrid-reinforced-earth embankment was about 1,000 m long, about 15-m-wide (at its top), and up to 4-m-high. The subject embankment was required to serve as an access road for construction equipment including heavy cranes, and also will be utilized as an access road for maintaining the new bridge during its service life. A subsurface investigation program including drilled borings was performed along the embankment alignment. The subsurface conditions typically consisted of surficial fill underlain by organic clay and peat overlying marine sand which is underlain by glacial till overlying bedrock. Figure 1 shows typical subsurface conditions encountered along the proposed embankment.

2.1 Embankment Configuration

Fill placement was typically required to construct the proposed embankment above existing grades. The height of the embankment ranged from about 1.5 m to about 4 m.

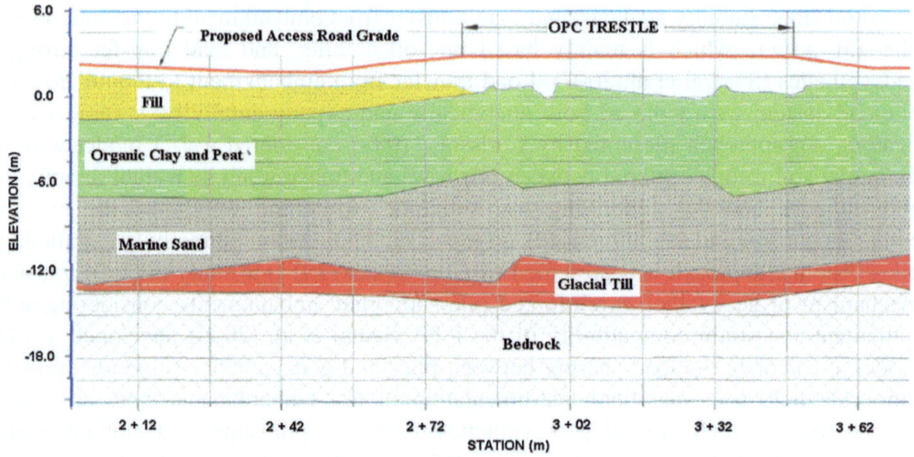

Fig. 1. Typical subsurface conditions along proposed embankment profile

Results of field and laboratory tests performed during geotechnical subsurface investigation indicated loose/soft nature of surficial fill and organic clay and peat. Loads anticipated on the proposed embankment from construction equipment including heavily loaded cranes were significantly high. A preliminary settlement analysis of a test embankment indicated proposed embankment, if supported on surficial fill and organic clay and peat soils, would experience settlements that would not be tolerable by the manufacturer of the heavily loaded cranes. Therefore, it was decided that a pile-supported structure will be necessary to reach acceptable performance of the proposed embankment. A conventional pile-supported load-relieving platform option was discussed. However, because of cost ineffectiveness considerations, this option was abandoned. Project environmental limitations precluded the use of steel, concrete, and aggregate material elements. Therefore, a geogrid-reinforced-earth embankment supported on driven timber piles was explored. The timber piles were specified as pressure-treated elements to ensure acceptable short- and long-term pile integrity. The purpose of the geogrid-reinforced-earth was to minimize piling cost by maximizing pile spacing as the geogrid-reinforced-earth mass was designed to act as a load-transfer mat to evenly distribute embankment loads on timber piles. The configuration of the proposed embankment section significantly changed along the embankment profile. The overall embankment footprint needed to be coordinated and minimized in order to limit the environmental impact on adjacent wetlands. Therefore, at some sections, a temporary vertical geogrid-reinforced-earth face had to be provided during the temporary use of the embankment, and then removed to form a permanent geogrid-reinforced earth slope, after the construction of the bridge is done (i.e. when the embankment will be used as an access road for bridge maintenance equipment). In addition, the top width of the embankment was increased at locations at which construction equipment was presumed to make turns. Figure 2 shows a typical section configuration of the proposed geogrid-reinforced-earth pile-supported embankment and Fig. 3 shows an aerial photo indicating the extent of the embankment after its construction.

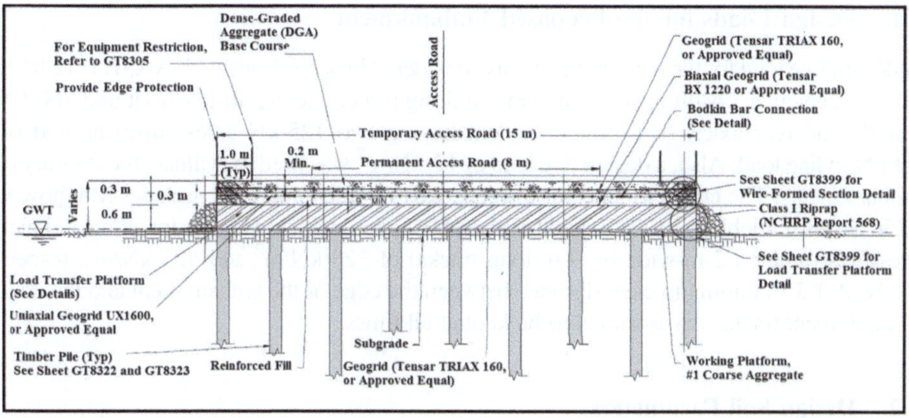

Fig. 2. Configuration of geogrid-reinforced-earth pile-supported embankment

Fig. 3. Aerial photo of the proposed geogrid-reinforced-earth pile-supported embankment after construction

3 Design Loads and Analysis of Proposed Embankment

In this section, design loads considered in the analysis and design of the proposed geogrid-reinforced pile-supported embankment are presented.

3.1 Design Loads for the Proposed Embankment

A design live load for the embankment was considered in accordance with AASHTO LRFD HL-93 Vehicular Loading, which includes a design truck having 40 kN front and 160 kN middle and rear axles, or a design tandem having two 125 kN axles superimposed on 10 kN/m line load. Also, a design live load of 12 kN/m^2 was used to evaluate the stability of embankments. The embankment was also checked for a live load that includes Manitowoc 2250 Series 3, Liebherr 1300SX, and Liebherr 895 cranes with maximum traveling and pick pressures (under 1.2-m-wide by 9-m-long tracks) of 325 kN/m^2 and 425 kN/m^2, respectively. A 1.5 m minimum clear distance between the edge of the embankment and the edge of equipment tracks was assumed to be kept at all times.

3.2 Design Soil Parameters

Field and laboratory tests were performed to estimate design soil parameters for different soil layers encountered during the subsurface investigation. Lab tests performed on soil samples extracted from the organic clay and peat included Atterburg limits, natural water content, strength (triaxial and torvane) and consolidation testing. For the organic clay and peat, index tests indicated natural water contents typically ranging from 90% to 110% and liquid limits typically ranging from 100% and 110%. Strength tests indicated the undrained shear strength of the organic clay/peat ranged from 5 to 25 kN/m^2. Because the shear strength and consolidation of the organic clay and peat layers control the behavior of the proposed embankment, it was decided to build and monitor a grade-supported test embankment to either confirm or modify strength and consolidation parameters from lab tests, based on field observations. The performance monitored for the test embankment, including embankment settlement and porewater pressure within organic clay and peat, agreed well with the embankment performance predicted based on soil parameters obtained from lab tests. The design soil parameters for the reinforced fill, existing fill, organic clay, peat, marine sand, and glacial till layers are given in Table 1.

3.3 Finite Element Analysis and Material Modeling

An initial design of the embankment was performed following the soil arching model presented by Carlsson (1987). A typical allowable axial compressive capacity of 300 kN per timber pile was used to evaluate pile spacing. Pile length was determined in such a way to keep minimum 3 m embedment into natural marine sand and/or glacial till soils. A finite element (FE) model was then built to simulate, and analyze the construction and the behavior of the embankment under various design loading scenarios. The analysis was performed using the commercial computer code Plaxis 2D. The embankment and underlying soils were idealized using 15-node, 2D plane-strain elements. The behavior of different soils was idealized as elasto-plastic materials with reinforced fill, existing fill, marine sand, and glacial till soils modeled as Mohr-Coulomb materials, and the organic clay and peat obeying the soft-soil-creep model. The timber piles were modeled as embedded pile elements and the geogrid reinforcement was modeled as elastic membrane. The FE model used to analyze the geogrid-reinforced-earth pile-supported embankment is shown in Fig. 4.

Table 1. Material models and design soil parameters

A - Strength and deformation parameters:						
Material	Unit weight (kN/m^3)	Soil model	*Phi* (Deg.)	S_u (kN/m^2)	E (kN/m^2)	v
Reinforced fill	19	Mohr-Coulomb	30	N/A	4×10^4	0.3
Existing fill	19	Mohr-Coulomb	30	N/A	2×10^4	0.3
Organic clay	14	Soft Soil Cr Model	N/A	7.5 to 17.5	N/A	0.45
Peat	12	Soft Soil Cr Model	N/A	12.5	N/A	0.45
Glacial till and marine sand	19	Mohr-Coulomb	30	N/A	6×10^4	0.35
B - Consolidation parameters:						
Material	C_c	C_r	K_x (m/s)	K_x/K_y	Remarks	
Organic clay	0.2	0.02	1.8×10^{-7}	5	Estimated based on data from monitored embankment section.	
Peat	0.3	0.03	1.8×10^{-7}	5		

Phi = Angle of Internal Shearing Resistance
S_u = Undrained Cohesion
E = Elasticity Modulus
v = Poisson's Ratio
C_c = Compression index
C_r = Re-compression index
Kx = Permeability in horizontal direction
Ky = Permeability in vertical direction

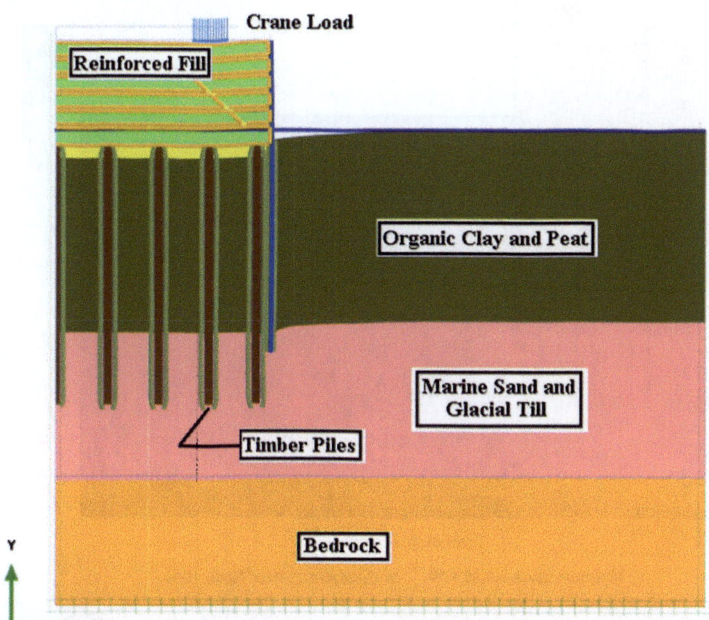

Fig. 4. FE model used to analyze the geogrid-reinforced-earth pile-supported embankment

3.4 Staged Construction and Stability Analysis of Proposed Embankment

The FE model described above was used to simulate the construction and operation of the proposed embankment using the Plaxis built-in staged-construction procedure. Construction stages simulated included installation of timber piles, building the initial load-transfer platform (LTP), building the geogrid-reinforced fill embankment and applying the embankment loads. After construction stages were simulated, the strength reduction method (SRM) was utilized to evaluate the embankment stability against different potential failure modes as defined by either local overstress or excessive deformation in the geogrid reinforcement, piles, and/or soft organic clay and peat soils due to loads directly transmitted to these elements.

3.5 Analysis Results and Design Case

Analysis results obtained from the FE model included deformations, stresses and forces. The results were obtained for each simulated construction stage and for the cases of the embankment being utilized as a temporary access road supporting heavy cranes, and as a permanent access road supporting bridge maintenance equipment. For the case of embankment under temporary crane loadings, Figs. 5 through 7 show results of vertical displacements of the entire FE mesh, effective vertical stresses on a horizontal plane immediately above heads of timber pile, and axial forces in the

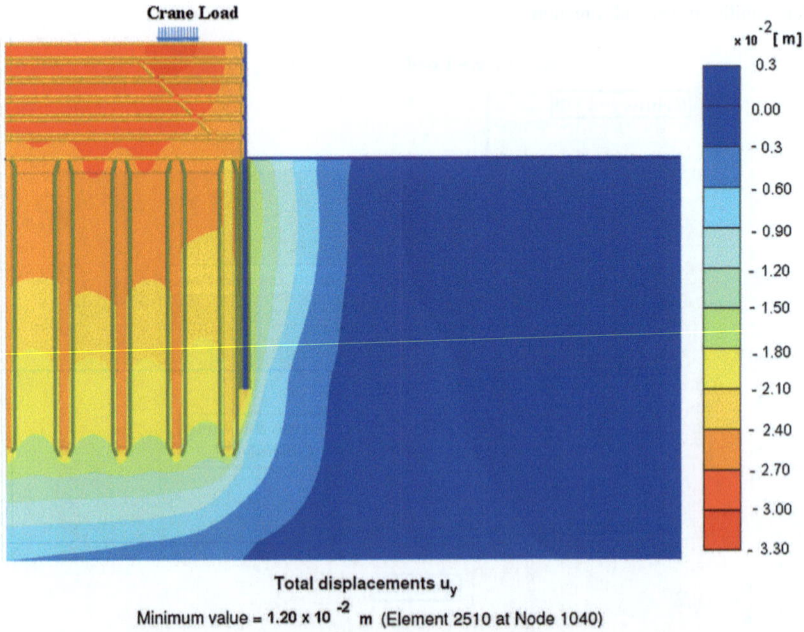

Fig. 5. Distribution of vertical displacement as predicted from FE analysis for the reinforced-earth embankment and underlying soils and piles

geogrid immediately above the timber piles. The FE analysis done with different pile spacing and different LTP geogrid stiffness indicated that the pile spacing as well as the stiffness of the geogrid reinforcement in the LTP (i.e. the number of geogrid layers x the single-layer stiffness) control the pressures imposed by the embankment on the underlying soil subgrade. Various runs were performed to reach a geogrid stiffness that resulted in minor pressure on the soil subgrade (i.e. maximum load shared by piles) while keeping pile spacing within 1.5 to 2.5 m and pile load within 300 kN per pile. The results of the FE analysis indicated the embankment behaves as a reinforced-earth mat transferring the majority of the embankment loads to the timber piles. This behavior is evident from the distribution of the vertical effective stresses (Fig. 6) on a horizontal plane immediately above timber pile heads. Such behavior resulted in minimizing embankment settlement and indicated successful embankment design.

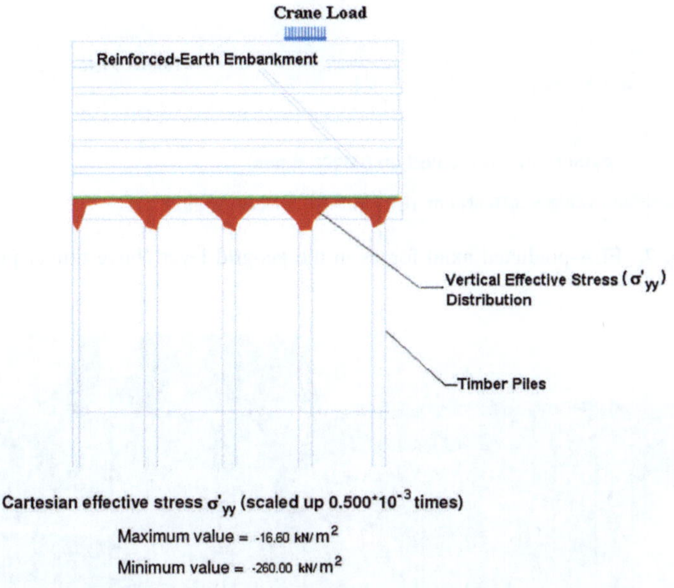

Fig. 6. FEA-predicted vertical effective stresses on horizontal plane immediately above timber piles

4 Embankment Construction and Performance

Figure 8 shows the timber pile installation for the proposed reinforced-earth pile-supported embankment. Because of the very soft nature of the organic clay and the peat encountered at some sections along the embankment profile, a layer of tri-axial geogrid with geotextile and granular fill was placed on top of the natural soil subgrade, to create a stable working platform for the pile installation equipment. After finalizing timber pile installation for each embankment segment, the geogrid-reinforced load-transfer platform was installed and the construction of the geogrid reinforced embankment was completed. The stiffness and strength of the geogrid was also designed in such a way to minimize load transfer to underlying soil subgrade.

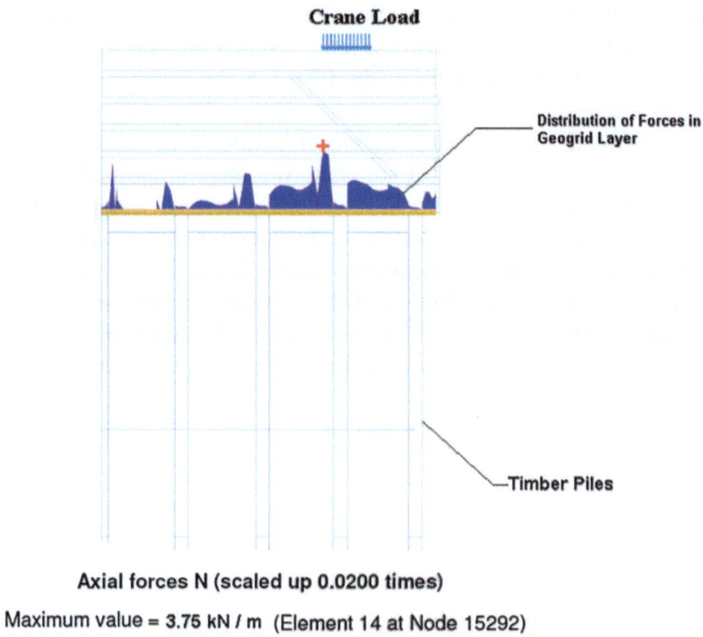

Fig. 7. FEA-predicted axial forces in the geogrid layer above timber piles

Fig. 8. Installation of timber piles during the construction of the geogrid-reinforced pile-supported embankment

Fig. 9. Geogrid-reinforced pile-supported embankment during operation

The critical case of loading resulted from the operation of the heavily loaded Liebherr 895 cranes while picking, see Fig. 9. The heavily loaded Liebherr 895 cranes were equipped with sensors to automatically stop crane operation in case the angular distortion resulting from ground differential settlement underneath any two points on the crane track exceeded a triggering limit. A timber mat was placed on top of the embankment to distribute the crane track load over the top of the embankment and to minimize differential settlement and angular distortion under the crane track. When subject to heavily loaded crane operations, the geogrid-reinforced-earth pile-supported embankment performed very well, and no differential settlement that required halting crane operation was observed under the crane track, as evidenced by the sensors attached to the crane.

5 Conclusions

The design and construction of a geogrid-reinforced pile-supported embankment was presented in this paper. Because of the complex behavior of the geogrid-reinforced embankment, the analysis of the embankment was performed utilizing finite element method. Subsurface investigation including field and lab testing was done before design starts. However, because of challenging geotechnical conditions encountered at the project site, a test embankment was built and monitored, and the monitoring results of the test embankment confirmed the design soil parameters obtained from lab tests. Results of numerical analysis indicated effectiveness of the proposed design of the geogrid-reinforced piled embankment. Settlements obtained from the numerical analysis indicated acceptable total and differential settlement of the embankment under heavily loaded crane operation, which agreed well with the embankment performance monitored during crane operation. Pile loads obtained based on the results of FEA indicated the Carlsson (1987) approach overestimated load share transmitted to piles, possibly because of Carlsson's approach ignores the effect of the geogrid stiffness and the soil-geogrid-pile interaction. The use of the FEA, rather than conventional analysis approaches, enabled performing full soil-geogrid-pile interaction and, hence, optimizing the geogrid stiffness. In addition, the use of the geogrid-reinforced LTP with optimized stiffness in combination with timber piles resulted in larger pile spacing and, hence, reduced the cost of the required piling.

References

Bhasi, A., Rajagopal, K.: Numerical modeling and analysis of geosynthetic reinforced pile supported embankments. In: Proceeding of the 5th WSEAS International Conference on Engineering Mechanics, UK, February 2012 (2012)

Byrum, C.R.: Analysis of pile supported embankments considering soil settlement and load transfer platform strain compatibility. Journal of the Transportation Research Board: Paper No. 15-2458 (2014)

Poulos, H.G.: Design charts for piles supporting embankments on soft clay. ASCE J. Geotech. Geoenviron. Eng. **133**(5), 493–501 (2007)

Briançon, L., Faucheux, G., Andromeda, J.: Full-scale experimental study of an embankment reinforced by geosynthetics and rigid piles over soft soil. In: Proceeding of EuroGeo, Scotland, September 2008 (2008)

Carlsson, B.: Reinforced Soil, Principles for Calculations: Terratema AB, LinkOping (in Swedish) (1987)

Navin, M.P.: Stability of embankments founded on soft soil improved with deep-mixing-method columns. Ph.D. dissertation, Virginia Polytechnic Institute and State University – The United States (2005)

van Eekelen, S.J.M., Bezuijen, A., van Tol, A.F.: Validation of analytical models for the design of basal reinforced piled embankments. Geotext. Geomembr. **43**, 56–81 (2015). www.elsevier.com/locate/geotexmem

Satibi, S., van der Meij, R., Leoni, M.: Piled embankments: literature review and required further research using numerical analysis. Institutsbericht 34, Institute for Geotechnical Engineering, University of Stuttgart, Germany (2007)

Svanø, G., Ilstad, T., Eiksund, G., Want, A.: Alternative calculation principle for design of piled embankments with base reinforcement. In: Proceeding of the 4th International Conference on Ground Improvement Geosystems, Helsinki, June, pp. 541–548 (2000)

Zhuang, Y., Cui, X.: Reinforced piled embankment for a high-speed railway over soft soil: a numerical and analytical investigation. Acta Geotech. Slov. **2**, 57–65 (2015)

Case Study on GeoTrel® Reinforced Earth® Steepened Slopes on Soft Founding Soils for the Approach of Major Bridge Over River Kaljani at CoochBehar in West Bengal, India

Sumonto Banerjee(✉), Atanu Adhikari, Saikat Chatterjee, and Dhananjoy Das

Reinforced Earth India Pvt. Ltd, New Delhi, India
sumonto.banerjee@terrearmeeindia.com

Abstract. The paper focuses on a particular case study involving the construction of full height steep reinforced slope walls with an inclination of 70° with the horizontal for the approach of major bridge over river Kaljani in West Bengal, India. The maximum height of the steep slope walls is 9m at the abutment location. The soft founding soil has been improved by soil replacement at a requisite design depth followed by providing inclusions of Geostrap® basal reinforcement at intermediate layers. The case study covers the design, engineering, construction methods and challenges that were overcome during the construction of this project. Another key feature in utilizing this technology over concrete panel or block facing was savings in construction time span and known as 'ready to install' technology. The project has saved the client time and money over conventional rigid retaining wall structures in such cases.

Keywords: GeoTrel® · Steep slope · Vertical wall · Flexible retaining structure · Soft soil · Mechanical connection

1 Introduction

GeoTrel® Reinforced Earth® retaining wall system is an innovative and cost effective solution for the construction of earth stabilized vertical walls and steep slopes. The facing system offers a natural aesthetic appearance that amalgamates well with the surrounding environment. These walls utilize a durable galvanized prefabricated steel mesh panel that adds more flexibility to the structure in order to accommodate differential settlement and dynamic loads. The GeoTrel® system is composed of geosynthetic reinforcing strips attached by patented connectors to a prefabricated and galvanized reinforcement mesh facing. The reinforcing strips are available in 2 types either with regular or high adherence edges, chosen according to the *pH* condition of the intended backfill:

- **GeoStrap®** reinforcing strip (coated polyester - PET) when the backfill has a *pH* lower than 9.

S.K. Shukla and E. Guler (eds.), *Advances in Reinforced Soil Structures*,
Sustainable Civil Infrastructures, DOI 10.1007/978-3-319-63570-5_12

- **EcoStrap™** reinforcing strip (coated polyvinyl alcohol - PVA) when the backfill has a *pH* higher than 9.

Design and construction of GeoTrel® shall include all dead and live loads and its combinations including seismic loads and necessary erosion protection (if required) for long term performance. The connectors between the strips and the facing have been designed to provide high durability and easy installation. The minimum diam of the reinforcement mesh bar is 8 mm having an average galvanization coating of 86 microns which provides high performance to the structure during the intended service life.

Before execution of works it is necessary to conduct requisite tests of the founding soil for checking the Safe Bearing Capacity in terms of overall stability and safety of the GeoTrel® slope structure.

The MSE walls are also very much earthquake resistance due to their flexibility and has resisted ground acceleration of 0.91 g (Sankey and Segrestin 2001) with minimum damage.

1.1 About the Project

The Public Works Department under Government of West Bengal intends to build one of the approaches of the prestigious bridge over river Kaljani but was facing few major challenges:

- Need to complete the approach within a shorter duration of 3 months, the reason being the next assembly election was scheduled at that time.
- The system has to be faster and economical as compared to conventional in-situ retaining wall system.
- Limited ROW due to existing town near the bridge site. Restoration would have taken few months followed by unfavorable situation that the department wish to avoid, as the election was nearby.
- Founding soil doesn't have adequate bearing capacity and needs ground improvement.
- Foundation treatment escalates the project cost further, depending on type, loading and behavior of the retaining structure.

Considering the above constraints, the department has approached us (Reinforced Earth India Pvt. Ltd.) and GeoTrel® reinforced earth® steepened slope has been considered as the ideal solution for the bridge approach.

Salient feature of the bridge approach

- Length of the approach – 250 m
- Maximum Height of retaining structure – 8 m (from existing ground level near abutment)
- Carriageway Width – 2 lane (11 m)
- Application – Road

- Traffic surcharge at top – 24 kPa
- Seismic Zone – V

Backfill Soil Properties

- Fill Density of Reinforced fill – 20 kN/m^3
- Fill Density of Retained fill – 20 kN/m^3
- Angle of Shear resistance of reinforced fill – 30°
- Angle of Shear resistance of retained fill – 30°

2 Design Criteria

The design concepts and criteria have been discussed by several agencies (USDT 2009; BS 2010; IS 2002; IRC 2014).

- **Principle of reinforced soil engineering**

A simple model helps to explain the principle on which the reinforced soil techniques are based. Let us consider a soil element (Fig. 1), which is part of an infinite mass of soil: the application of a vertical stress σ_v produces a deformation in the element and the consequent horizontal stress σ_h, generated by the lateral compression suffered by the adjacent soil. Horizontally, the soil element undergoes a "tensile deformation" e_h, which is one of the main causes of local failures.

Fig. 1. 3D View of a GeoTrel® Structure

When a reinforcing element is put into the soil, the application of a vertical stress is followed by the deformation of the soil element and the extension of the reinforcement. This extension then generates a tensile strength F in the reinforcement, which in turn produces a horizontal stress σ_h^*. This stress, which also provides a confinement action of the soil granules, greatly contributes to resist the horizontal forces and to reduce the horizontal deformations. Therefore, the inclusion of a soil reinforcement into the soil mass reduces the stresses and strains applied to the soil; on the other hand the vertical stress σ_v applied to the soil mass can be increased, compared to the un reinforced soil, at equal deformations.

- **Properties of the soil reinforcement** (Table 1)
  (a) **Friction behavior** of the polymeric strap is defined by the pullout resistance which is defined by the ultimate tensile load required to generate outward sliding of the reinforcement through the reinforced soil mass.
  (b) **Ultimate tensile strength** of Polymeric strip is defined as force required to rupture the reinforcement. The ranges of Polymeric strip available have ultimate tensile strengths varying from 25 kN to 50 kN
  (c) **Creep behavior**
  (d) **Installation damage**
  (e) **Long term durability**

  The summary of all reduction factors are as follows:

$$\text{LTDS} = Tult \times (RF_{CR} \times RF_{ID} \times RF_D)/FS_R$$

RF_{CR} = 0.69 (Reduction factor for creep)
RF_{ID} = 0.95 (Reduction factor for Installation Damage)
RF_D = 0.87 (Reduction factor for Durability)

The apparent coefficient of friction μ^*, between the fill and the polymeric strip reinforcement can be conservatively considered as 1.1 at top and linearly varying to 0.8tanΦ at a depth of 6 m from top and down. For the reinforced earth fill specification used on this project (ϕ' = 30^0 Coefficient of Uniformity $C_U \geq$ 2), μ^* varies from a value of 1.100 at a depth of 0.0 m to 0.50 at depth of 6 m or greater.

- **Backfill requirement** (Table 2)
  The design of high embankment is based on the following criteria of fill soil:
  1. The fill in the structure or slope shall be wholly frictional.
  2. The fill for reinforced soil structures shall be well graded selected material as specified and available within reasonable lead distance. The fill must allow

Table 1. Technical specification of POLYMERIC GeoStrap®

Properties	Unit	Minimum average roll value		
Ultimate tensile strength	kN	25	37.5	50
Width	mm	49	49	49
Long term design strength	kN	14.26	21.39	28.51
Considering ramification *(considering safety factor of 1.1)*	kN	12.96	19.44	25.2

Table 2. Mechanical requirements of RE Fill

Sieve size	Percent passing
80 mm (gravel)	100%
4.75 mm (coarse sand)	More than 75%
75 micron (silt)	Less than 15%

dissipation of pore pressure by designing the same with free draining characteristics and by providing vertical and horizontal drainage provisions in the reinforced soil volume. The association of drainage bay or interface drains shall be connected properly to the gradient required and shall be maintained during compaction in layers.

Acceptance limits for materials with more than 15% passing 75 micron are related to the percentage of particles smaller than 15 microns as follows:

- Materials with more than 15% passing 75 **micron** sieve and more than 20% of particles smaller than 15 microns are inadequate and shall not be used except
- Materials with more than 15% passing 75 micron sieve and 10% of particles smaller than 15 microns are acceptable provided that the internal friction angle is not smaller than 32° and the fill shall be non-plastic.

- **Global stability**

Failure surfaces passing through the reinforced soil block and extending beyond it or encompassing the foundation soil must be analyzed to ensure sufficient length and strength within the reinforced soil mass (see Fig. 2).

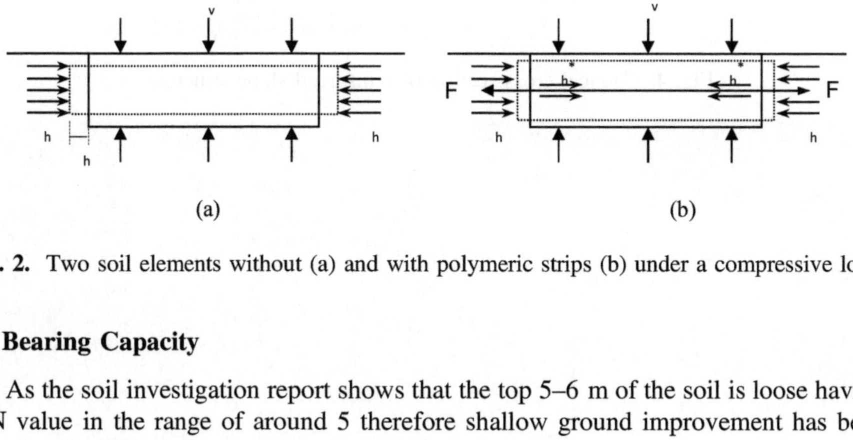

Fig. 2. Two soil elements without (a) and with polymeric strips (b) under a compressive load.

- **Bearing Capacity**

As the soil investigation report shows that the top 5–6 m of the soil is loose having a N value in the range of around 5 therefore shallow ground improvement has been done in the following way (Figs. 3, 4, 5, 6, 7, 8 and 9):

(a) Upto 5 m height of reinforced slope: Only soil replacement of 1 m below PCC Pad Level.

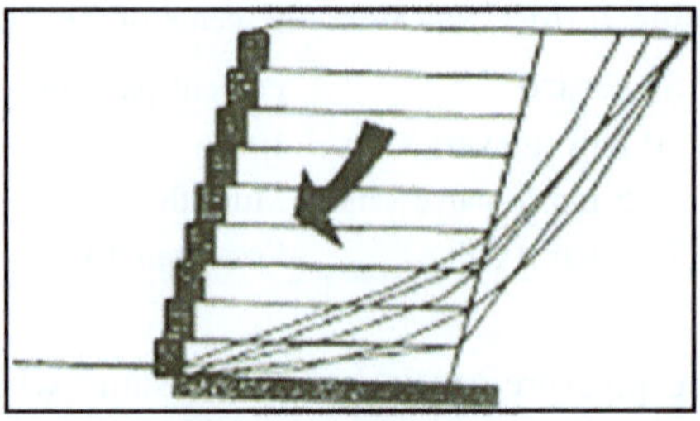

Fig. 3. Global Failure Mode – Example of failure surfaces (polygonal or circular) passing through the reinforced block and extending to soil behind it

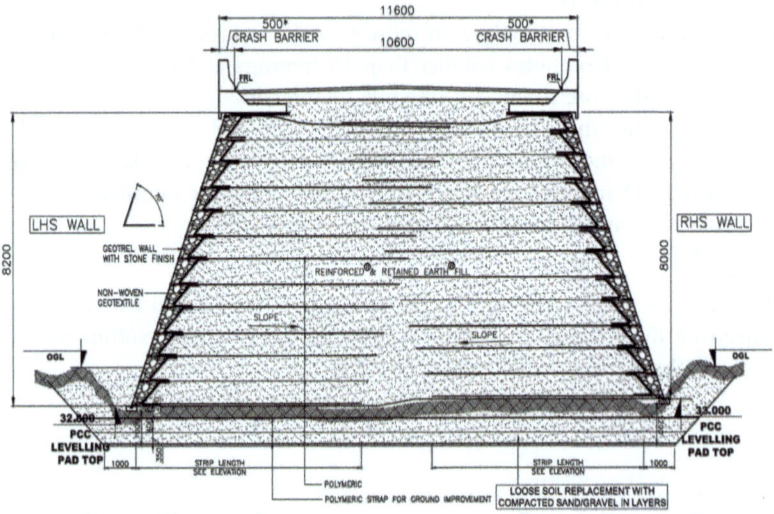

Fig. 4. Typical cross-section of reinforced slope structure

Fig. 5. Erection of first layer of Geotrel

Fig. 6. Erection & boulder filling completed for first layer

Fig. 7. Erection of Geotrel in progress

Fig. 8. Compaction of Reinforced Earth fill in progress

Fig. 9. Constrcution of Geotrel wall completed

(b) From 5 m height to rest of the reinforced slope: 25 kN Grade Geostrap having 20 m length has been laid in lateral direction @ 0.25 m horizontal spacing in 3 layers @ average vertical spacing of 0.3 m.

- **Seismic Analysis**

Horizontal seismic acceleration coefficient, Ao represents the maximum horizontal ground acceleration for a particular area. Recognizing the conservative nature of pseudostatic analysis, FHWA recommends using design seismic acceleration, Am, which is half the maximum value (see Geotechnical Engineering Circular No. 3, Design Guidance: Geotechnical Earthquake Engineering For Highways, Vol. I & II, May 1997, Publication No. FHWA-SA-97-077). Also reference can be made to Page 241 of FHWA-NHI-10-025 for design seismic acceleration. It may also be noted that many Reinforced Earth™ structures have been designed to this standard and subjected to significant seismic events such as the devastating earthquake, which struck Turkey in August 1999. This 45-second earthquake, measuring 7.4 on the Richter scale, caused widespread damage to buildings and significant loss of life. Whilst several highway structures suffered irreparable damage, the Reinforced Earth™ structures remained largely serviceable. This confirms the stability of such flexible structures against earth quake.

The value of maximum ground acceleration (α_o) largely depends on which of the four classified zones of the country the structure is located within. The design seismic acceleration considered is 0.36 g (IS 1893: 1984) for Zone–V as per IS 1893: 2002, Part-1. This has been interpreted as a horizontal bedrock acceleration of 0.18 g. It has been assumed that the seismic design case is to be considered as an accidental load case and that the effect of traffic surcharge is reduced to 50% during the seismic event. The long term design strength for GeoStrap® reinforcement do not require a creep reduction factor for the seismic loading condition. The dynamic component of load for seismic design is a transient load and does not cause strength loss due to creep as per Sect. 7.1.2.a of FHWA-NHI-10

- **Drainage system**

The drainage system is the most important function for any slope construction. However, due to the design considerations for frictional properties of the reinforced volume is relatively permeable, in many circumstances this reinforced mass is as effective as a drain without the use of other means. But, considering the large height and width of the embankment it is very important to have proper drainage system to improve the overall stability of the structure.

The drainage system consists of 450 mm wide crushed stone of size 150 mm to 200 mm provided at the facia throughout the entire height of reinforced slope structure followed by a layer of non-woven geotextile.

Surface drainage system: Surface water runoff drainage system shall be adopted above the reinforced slope and channeled with suitable slope to drain out the water to a nearest outlet. Longitudinal lined surface drain shall be provided at the toe of each slope.

3 Construction Method

Construction of reinforced earth® steepened slope is a simple and repetitive method which is mostly labour oriented and have lesser usage of machineries or construction equipments. The construction steps are:

- The first row of panels (i.e., galvanized prefabricated steel mesh facia) is installed on a well-levelled and compacted technical fill with requisite inward batter/slope (which is 70 degree w.r.t horizontal in this project).
- The first row of panel is braced directly to the ground to prevent movement during placement and compaction of selected backfill in layers.
- The succeeding panel courses are installed as the geotextile backing, backfill, crushed stone packing at the fascia and GeoStraps (soil reinforcing elements) are placed.
- Once installed, each layer of Geostraps is vertically spaced 600 mm apart, which corresponds to a multiple of the backfill layer thickness.
- The backfill is placed and compacted using vibratory rollers, except near the fascia. The 1.5 m width from the fascia should be compacted by means of plate compactor or a 1Ton baby roller.
- The above methods were followed in sequence to reach at top followed by pavement works and road furniture and fixtures.

4 Conclusions

Recognized as a major innovation in the field of civil engineering, the reinforced earth® steepened technique provides numerous structural solutions for owners and contractors ranging from retaining walls to bridge abutments and have following significant advantages:

- Lower global cost: the possibility to build steeper slopes reduces the quantity of fill material needed for an embankment;
- Use of more natural resources and relatively less quantity of manufactured product and hence less emulsion of carbon dioxide (CO_2) and less impact on global warming;
- Relatively less use of equipment for construction and hence again less impact on global warming, besides saving in direct cost
- Improved stability: the reinforcement allows to increase the factor of safety;
- Reinforced soil structure is inherently flexible, it is possible to build directly on a foundation soil with low bearing capacity; a reinforcement at the base allows to build on soft soils, which would usually require a preliminary consolidation and great caution during construction.
- High resistance against earthquake load due to flexible in nature.
- Use of special facing units will give an even face for good appearance
- It encourages the creation of an environment that provides support for plant and animal life.
- The construction blends easily into the landscape and is difficult to identify as reinforced earth structures after a few years.

References

U.S Department of Transportation (USDT): Design & construction of mechanically stabilized earth walls & reinforced soil slopes. In: FHWA-GEC 011 vols. I and II (2009)

British Standards (BS): Code of practice for strengthened-reinforced soils and other fills. BS 8006-1:2010 (2010)

Indian Standards (IS): Criteria for Earth Resistant Design of Structures. IS 1893:2002, Fifth Revision (2002)

Indian Road Congress (IRC): Guidelines for Design & Construction of Reinforced Soil Wall. IRC:SP:102-2014 (2014)

Sankey, J.E., Segrestin, P.: Evaluation of seismic performance in MSE structures. In: Proceedings of the International Symposium on Earth Reinforcement. Landmarks in Earth Reinforcement, vol. 1, Kyushu, Japan, pp. 449–452 (2001)

Prediction of Ultimate Bearing Capacity of Eccentrically Loaded Rectangular Foundations Using ANN

B.P. Sethy[1(✉)], C.R. Patra[1], N. Sivakugan[2], and B.M. Das[3]

[1] National Institute of Technology Rourkela, Rourkela, India
barada.jeetu@gmail.com, crpatral9@yahoo.co.in
[2] James Cook University, Townsville, Australia
siva.sivakugan@jcu.edu.in
[3] California State University, Sacramento, USA
brajamdas@gmail.com

Abstract. Extensive laboratory model tests were conducted on a rectangular embedded foundations resting over homogeneous sand bed and subjected to eccentric load to determine the ultimate bearing capacity. The depth of embedment varies from 0 to *B* with an increment of 0.5*B*; where *B* is the width of foundation and the eccentricity ratio (*e*/*B*) was varied from 0 to 0.15 with increments of 0.05. Based on the laboratory model test results, a neural network model has been developed to estimate the reduction factor (*RF*). The reduction factor can be used to estimate the ultimate bearing capacity of an eccentrically loaded foundation from the ultimate bearing capacity of a centrally loaded foundation. A thorough sensitivity analysis has been carried out to determine the important parameters affecting the reduction factor. Importance was given on the construction of neural interpretation diagram, and based on this diagram, whether direct or inverse relationships exist between the input and output parameters was determined. The results from artificial neural network (ANN) were compared with the laboratory model test results and the agreement is good.

Keywords: Eccentric load · Rectangular foundation · Depth of embedment · Sand · Neural network · Reduction factor

1 Introduction

During the last thirty years, a number of laboratory model test results and a few field test results have been published that are related to the ultimate bearing capacity of shallow foundation resting over homogeneous soil. Most of the experimental studies relates to condition of centric loading. However, none of the published studies address the effect of load eccentricity on the ultimate bearing capacity of rectangular foundation using ANN. The purpose of this study is to develop a neural network model from the results of laboratory model tests to estimate the reduction factor. Artificial neural network (ANN) is an artificial intelligence system inspired by the behavior of human brain and nervous system. In the present study a feed forward back propagation neural network model has been used to predict the reduction factor of eccentrically loaded

S.K. Shukla and E. Guler (eds.), *Advances in Reinforced Soil Structures*,
Sustainable Civil Infrastructures, DOI 10.1007/978-3-319-63570-5_13

rectangular foundation. Backpropagation neural network is most suitable for prediction problems and Levenberg-Mar quadrt algorithm is adopted as it is efficient in comparison to gradient descent backpropagation algorithm (Goh *et al.* 2005; Hornik *et al.* 1989). By drawing a neural interpretation diagram relationship in between input parameters and output are found out. A prediction model is developed based on the weights of the ANN model. The developed reduction factor is compared with the experimental reduction factor.

2 Analysis and Data

All the laboratory model tests were conducted using a poorly graded sand with effective grain size $D_{10} = 0.325$ mm, uniformity coefficient $C_u = 1.45$, and coefficient of gradation $C_c = 1.15$. Model foundations used for the tests had dimensions of 100 mm 100 mm ($B/L = 1$), 100 mm × 200 mm ($B/L = 0.5$), 100 mm × 300 mm ($B/L = 0.33$) and 100 mm × 500 mm ($B/L \approx 0$). Steel plates having thickness of 30-mm were used to make the model foundations. The bottom of the foundation was made rough by applying glue and rolling the steel plate over sand.

Forty eight laboratory model tests were conducted. Three parameters e/B, B/L and D_f/B are used as inputs in the ANN model, and the output is the reduction factor RF given by

$$RF = \frac{q_{u(B/L,\, D_f/B,\, e/B)}}{q_{u(B/L,\, D_f/B,\, e/B=0)}} \quad (1)$$

where $q_{u(B/L,\, D_f/B,\, e/B)}$ is the ultimate bearing capacity with eccentricity ratio e/B and B/L ratio and an embedment ratio D_f/B; and $q_{u(B/L,\, D_f/B,\, e/B=0)}$ is the ultimate bearing capacity with centric vertical loading ($e/B = 0$) with B/L ratio at an embedment ratio D_f/B.

Out of 48 tests, 36 tests are considered for training and the remaining 12 are considered for testing. All the inputs and output are normalized in the range of [−1, 1] before training. A feed-forward back-propagation neural network is used with hyperbolic tangent sigmoid function and linear function as the transfer function. The network is trained with Levenberg-Marquardt (LM) algorithm as it is efficient in comparison to gradient descent back-propagation algorithm. The ANN has been implemented using MATLAB V 7.11.0 (R2015b).

3 Results and Discussion

Three inputs and one output parameters were considered in the ANN model. The schematic diagram of the ANN architecture is shown in Fig. 1. which was computed from the database. The number of neurons in hidden layer is varied and the optimum number was taken based on mean square error (mse) value which was maintained at 0.001. In this ANN model there were six neurons evaluated in hidden layer as shown in Fig. 2. Therefore the final ANN architecture as 3-6-1[i.e. 3 (input) – 6 (hidden layer neuron) – 1 (output)].

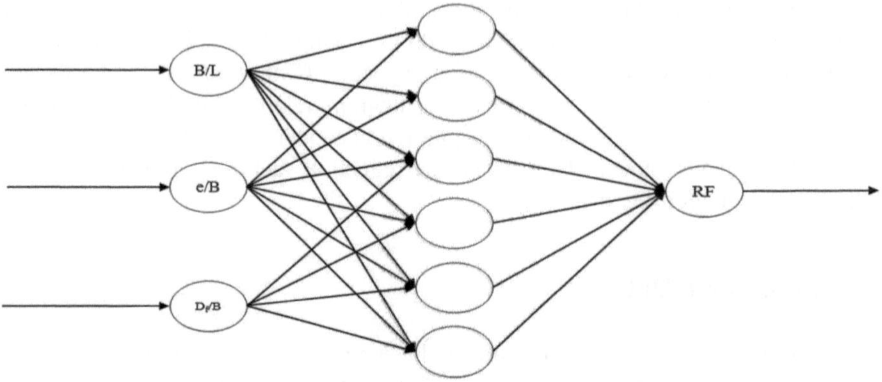

Fig. 1. ANN architecture

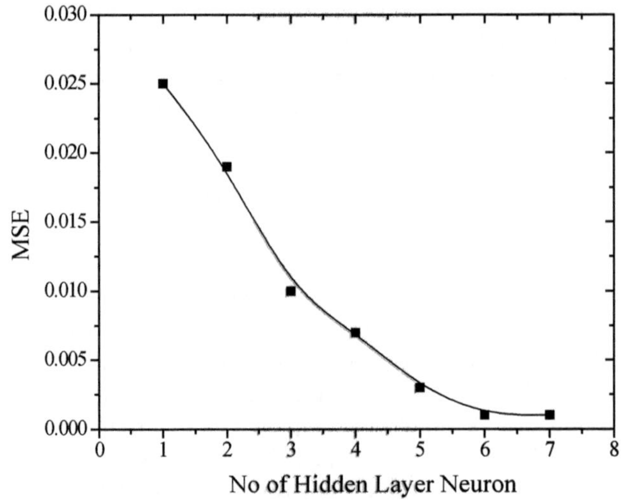

Fig. 2. Variation of hidden layer neuron with mean square error (mse)

Mean square error (MSE) is defined as

$$MSE = \frac{\sum_{i=1}^{n} \left(RF_i - RF_p\right)^2}{n} \quad (2)$$

Coefficient of efficiency, R^2 is defined as

$$R^2 = \frac{E_1 - E_2}{E_1} \quad (3)$$

where,

$$E_1 = \sum_{i=1}^{n} \left(RF_i - \overline{RF}\right)^2 \tag{4}$$

and

$$E_2 = \sum_{i=1}^{n} \left(RF_p - RF_i\right)^2 \tag{5}$$

where, RF_i, $\overline{RF}$ and RF_p are the experimental, average experimental, predicted RF values respectively; and n = number of training data.

The coefficient of efficiency (R^2) is found to be 0.995 for training and 0.902 for testing as shown in Figs. 3 and 4. The weights and biases of the network are presented in Table 3. These weights and biases can be utilized for interpretation of relationship in between the inputs and output, sensitivity analysis and framing an ANN model in the form of an equation. The residual analysis was carried out by calculating the residuals in between experimental reduction factor and predicted reduction factor for training data. Residuals can be defined as the difference between the experimental and predicted RF value and is given by

$$e_r = RF_i - RF_p \tag{6}$$

The residuals are plotted with the experimental number as shown in Fig. 5. It is observed that the residuals are evenly distributed along the horizontal axis of the plot.

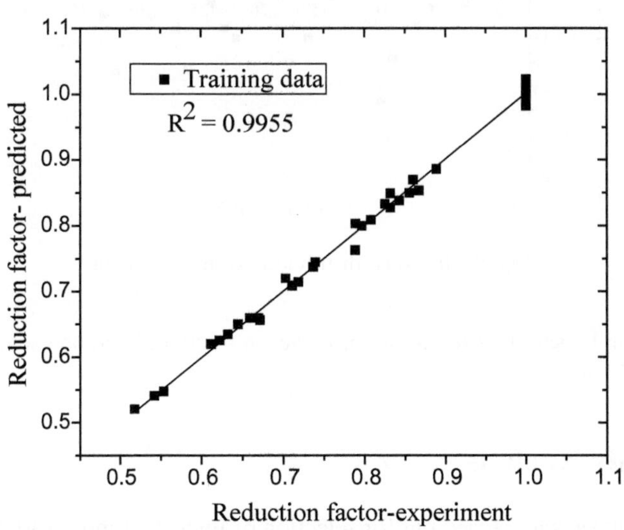

Fig. 3. Correlation between prediction reduction factors with experimental reduction factor for training data

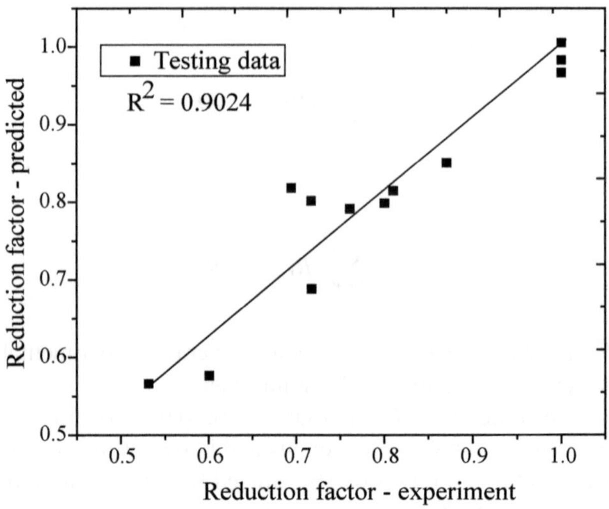

Fig. 4. Correlation between prediction reduction factors with experimental reduction factor for testing data

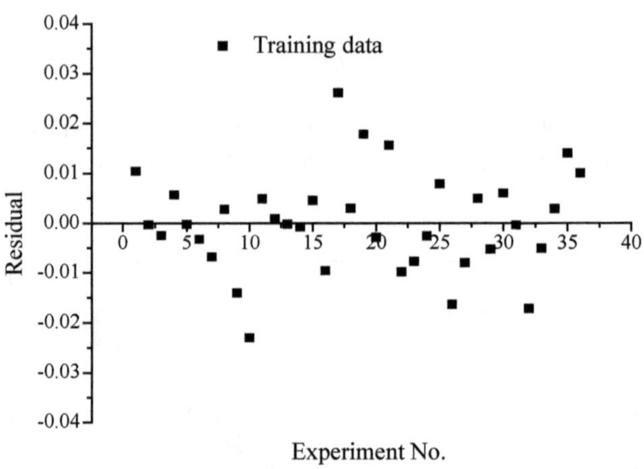

Fig. 5. Residual distribution of training data

Therefore it can be said that the network is well trained and can be used for prediction with reasonable accuracy.

4 Sensitivity Analysis

Sensitivity analysis was carried out for selection of important input variables. Different approaches have been suggested to select the important input variables. The Pearson correlation coefficient is one of them in selecting proper inputs for the ANN model. It was approached by Guyon and Elisseff (2003); Wilby *et al.* (2003). Goh (1994); Shahin *et al.* (2002); Behera *et al.* (2013) have used Garson's algorithm (Garson 1991) in which the input-hidden and hidden-output weights of trained ANN model are partitioned and the absolute values of weights are taken to select the important input variables. It does not provide information on the effect of input variables in terms of direct or inverse relation to the output. Olden *et al.* (2004) proposed a connection weight approach based on the neural interpretation diagram (NID), in which the actual values of input-hidden and hidden-output weights are taken. Table 4 shows the cross-correlation of the three input parameters with the reduction factor (*RF*) value. From the table it can be seen that *RF* is highly correlated to *e*/*B* with a values of 0.975 followed by D_f/B and *B*/*L*. The relative importance, *quantified through the parameter* S_i of three input parameters as per Garson's algorithm is presented in Table 5. The *e*/*B* is found to be the most important input parameters with relative importance value being 45.08% followed by 36.41% for *B*/*L* and 18.51% for D_f/B. As per the connection weight approach (Olden *et al.* 2004) the relative importance of the present input variables is also presented in Table 5. *B*/*L* is the most important input parameter (S_i = 8.6) followed by D_f/B (S_i = 1.38) and *e*/*B* (S_i = −1.06). The S_i values being positive imply that both *B*/*L* and D_f/B are directly related and *e*/*B* is indirectly related to *RF*. In other words increase in *B*/*L* or D_f/B leads to increase in *RF* and leads to increase in ultimate bearing capacity. Increasing *e*/*B* decreases the *RF*, and hence decreases the ultimate bearing capacity.

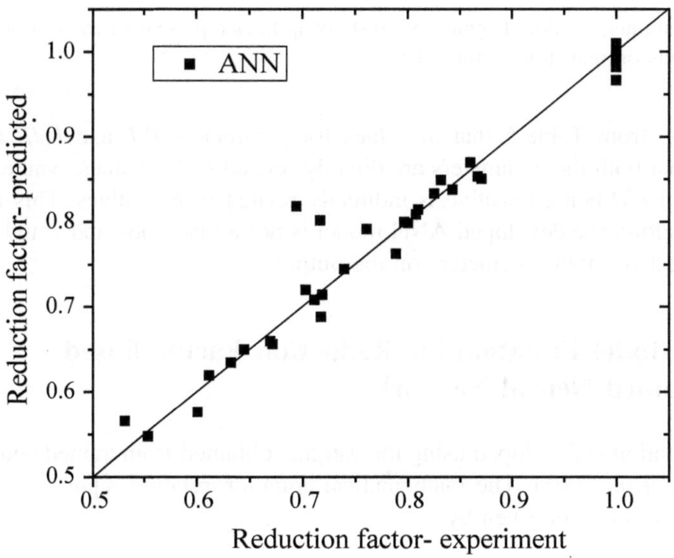

Fig. 6. Comparison of ANN results with experimental RF

5 Neural Interpretation Diagram (NID)

Ozesmi and Ozesmi (1999) proposed neural interpretation diagram for visual interpretation of the connection weight among the neurons. For the present study with the weights as obtained and shown in Table 3, an NID is presented in Fig. 7. The lines joining the input-hidden and hidden output neurons represent the weights. The positive weights are represented by solid lines and negative weights by dashed lines and the thickness of the line is proportional to its magnitude.

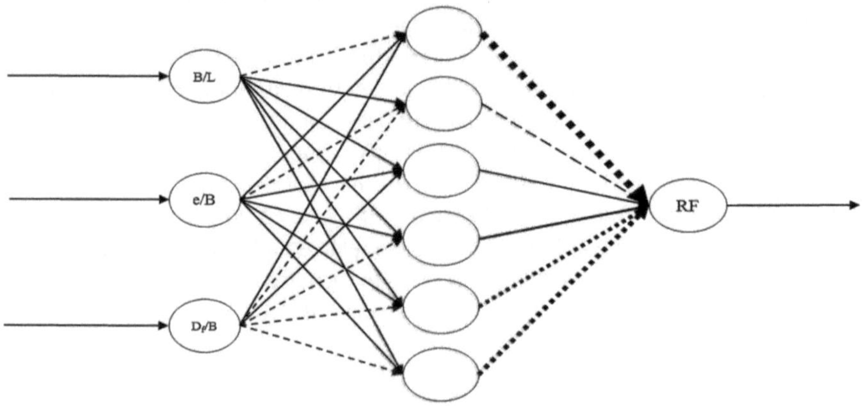

Fig. 7. Neural interpretation diagram (NID) showing lines representing connection weights and effects of inputs on reduction factor (RF)

It is seen from Table 5 that S_i values for parameters *B*/*L* and D_f/B are positive indicating that both the parameters are directly related to *RF* values, whereas S_i values for parameter *e*/*B* being negative is indirectly related to *RF* values. This is shown in Fig. 7. Therefore, the developed ANN model is not a black box and could explain the physical effect of input parameters on the output.

6 ANN Model Equation for Reduction Factor Based on Trained Neural Network

A model equation is developed using the weights obtained from trained neural network model (Goh *et al.* 2005). The mathematical equation relating input parameters (*B*/*L*, *e*/*B* and D_f/B) to output given by

$$RF_n = f_n\left\{b_0 + \sum_{k=1}^{h}\left[w_k f_n\left(b_{hk} + \sum_{i=1}^{m} w_{ik} X_i\right)\right]\right\} \tag{7}$$

where RF_n is the normalized value of *RF* in the range [−1, 1], f_n is the transfer function, *h* is the number of neurons in the hidden layer, X_i is the normalized value of inputs in the range [−1, 1], m is the number of input variables, w_{ik} is the connection weight between the ith layer of input and k^{th} neuron of hidden layer, w_k is the connection weight between the kth neuron of hidden layer and single output neuron, b_{hk} is the bias at the kth neuron of hidden layer and b_0 is the bias at the output layer.

The model equation of RF of shallow rectangular foundations subjected to eccentrically inclined load was formulated using the values of the weights and biases shown in Table 3 as per the following steps.

- Step 1

The input parameters were normalized in the range [−1, 1] by the following expressions

$$X_n = 2\left(\frac{X_n - X_{\min}}{X_{\max} - X_{\min}}\right) \tag{8}$$

- Step 2

Calculate the normalized value of reduction factor (RF_n) using the following expressions

$$A_1 = -0.0679\left(\frac{B}{L}\right)_n + 0.9077\left(\frac{e}{B}\right)_n + 0.0742\left(\frac{D_f}{B}\right)_n + 2.1 \tag{9}$$

$$A_2 = 11.43\left(\frac{B}{L}\right)_n - 18.11\left(\frac{e}{B}\right)_n - 0.95\left(\frac{D_f}{B}\right)_n + 20.89 \tag{10}$$

$$A_3 = 24.94\left(\frac{B}{L}\right)_n + 15.28\left(\frac{e}{B}\right)_n + 13.52\left(\frac{D_f}{B}\right)_n + 3.88 \tag{11}$$

$$A_1 = 26.69\left(\frac{B}{L}\right)_n + 1.16\left(\frac{e}{B}\right)_n - 14.61\left(\frac{D_f}{B}\right)_n + 10.28 \tag{12}$$

$$A_1 = 0.56\left(\frac{B}{L}\right)_n + 2.18\left(\frac{e}{B}\right)_n 0.83\left(\frac{D_f}{B}\right)_n - 1.86 \tag{13}$$

$$A_1 = 1.13\left(\frac{B}{L}\right)_n + 0.74\left(\frac{e}{B}\right)_n - 0.41\left(\frac{D_f}{B}\right)_n + 0.94 \tag{14}$$

$$B_1 = -4.36\left(\frac{e^{A_1} - e^{-A_1}}{e^{A_1} + e^{-A_1}}\right) \tag{15}$$

$$B_2 = -0.11\left(\frac{e^{A_1} - e^{-A_1}}{e^{A_1} + e^{-A_1}}\right) \tag{16}$$

$$B_3 = 0.14\left(\frac{e^{A_1} - e^{-A_1}}{e^{A_1} + e^{-A_1}}\right) \tag{17}$$

$$B_4 = -0.26\left(\frac{e^{A_1} - e^{-A_1}}{e^{A_1} + e^{-A_1}}\right) \tag{18}$$

$$B_5 = -0.52\left(\frac{e^{A_1} - e^{-A_1}}{e^{A_1} + e^{-A_1}}\right) \tag{19}$$

$$B_6 = -0.63\left(\frac{e^{A_1} - e^{-A_1}}{e^{A_1} + e^{-A_1}}\right) \tag{20}$$

$$C_1 = B_1 + B_2 + B_3 + B_4 + B_5 + B_6 + 4.27 \tag{21}$$

$$RF_n = C_1 \tag{22}$$

- Step 3

Denormalize the RF_n value obtained from Eq. 22 to actual RF as

$$RF = 0.5(RF_n + 1)(RF_{\max} - RF_{\min}) + RF_{\min} \tag{23}$$

$$RF = 0.5(RF_n + 1)(1 - 0.52) + 0.52 \tag{24}$$

Figure 6 shows the comparison of reduction factor obtained from Eqs. 1 and 23. It can be seen that the ANN results are closer to the experimental value. The deviation between the experimental and predicted RF is within ±10% except two values as shown in Table 1. The proposed ANN model can be used as an effective tool in predicting the RF and hence, the ultimate bearing capacity of an eccentrically loaded rectangular footing.

Table 1. Database used for ANN model and compared with experimental results

Data type	Test no.	B/L	e/B	D_f/B	Experimental q_u (kN/m^2)	$RF_{expt.}$	RF_{ANN}	Deviation (%)
Training	1	0	0	0	166.67	1.00	1.00	0.00
	2	0	0.1	0	109.87	0.66	0.66	−0.04
	3	0	0.15	0	86.33	0.52	0.52	−0.49
	4	0	0.05	0.5	226.61	0.86	0.85	0.66
	5	0	0.1	0.5	195.22	0.74	0.74	−0.04
	6	0	0.15	0.5	164.81	0.62	0.63	−0.53
	7	0	0	1	353.16	1.00	1.01	−0.68
	8	0	0.05	1	313.92	0.89	0.89	0.31

(continued)

Table 1. (*continued*)

Data type	Test no.	*B*/*L*	*e*/*B*	D_f/*B*	Experimental q_u (kN/m^2)	$RF_{expt.}$	RF_{ANN}	Deviation (%)
	9	0	0.1	1	278.6	0.79	0.80	−1.78
	10	0.33	0	0	131	1.00	1.02	−2.30
	11	0.33	0.05	0	109	0.83	0.83	0.58
	12	0.33	0.15	0	71	0.54	0.54	0.16
	13	0.33	0	0.5	224	1.00	1.00	−0.02
	14	0.33	0.1	0.5	181	0.81	0.81	−0.09
	15	0.33	0.15	0.5	161	0.72	0.71	0.63
	16	0.33	0.05	1	289	0.86	0.87	−1.11
	17	0.33	0.1	1	265	0.79	0.76	3.31
	18	0.33	0.15	1	239	0.71	0.71	0.42
	19	0.5	0	0	128	1.00	0.98	1.78
	20	0.5	0.05	0	102	0.80	0.80	−0.37
	21	0.5	0.1	0	86	0.67	0.66	2.32
	22	0.5	0	0.5	212	1.00	1.01	−0.98
	23	0.5	0.05	0.5	175	0.83	0.83	−0.94
	24	0.5	0.15	0.5	134	0.63	0.63	−0.42
	25	0.5	0	1	327	1.00	0.99	0.79
	26	0.5	0.1	1	230	0.70	0.72	−2.32
	27	0.5	0.15	1	200	0.61	0.62	−1.30
	28	1	0.05	0	102	0.84	0.84	0.59
	29	1	0.1	0	78	0.64	0.65	−0.82
	30	1	0.15	0	67	0.55	0.55	1.09
	31	1	0	0.5	238	1.00	1.00	−0.04
	32	1	0.05	0.5	198	0.83	0.85	−2.06
	33	1	0.1	0.5	176	0.74	0.74	−0.69
	34	1	0	1	339	1.00	1.00	0.29
	35	1	0.05	1	294	0.87	0.85	1.62
	36	1	0.15	1	227	0.67	0.66	1.50
Testing	37	0	0.05	0	133.42	0.80	0.80	0.24
	38	0	0	0.5	264.87	1.00	1.01	−0.52
	39	0	0.15	1	245.25	0.69	0.82	−17.82
	40	0.33	0.1	0	94	0.72	0.69	4.06
	41	0.33	0.05	0.5	195	0.87	0.85	2.29
	42	0.33	0	1	336	1.00	0.97	3.31
	43	0.5	0.15	0	68	0.53	0.57	−6.50
	44	0.5	0.1	0.5	152	0.72	0.80	−11.80
	45	0.5	0.05	1	265	0.81	0.81	−0.51
	46	1	0	0	121	1.00	0.98	1.70
	47	1	0.15	0.5	143	0.60	0.58	4.07
	48	1	0.1	1	258	0.76	0.79	−3.99

Table 2. Statistical values of the parameters

Parameters	Maximum value	Minimum value	Average value	Standard deviation
e/B	0.15	0	0.075	0.056
B/L	1	0	0.46	0.36
D_f/B	1	0	0.5	0.41
RF	1	0.52	0.8	0.15

Table 3. Values of connection weights and biases

Neuron		Weight				
		w_{ik}		w_k	Bias	
	B/L	*e/B*	*D_f/B*	*RF*	b_{hk}	b_0
Hidden neuron 1 (k = 1)	−0.0679	0.9077	0.0742	−4.3646	2.1037	4.2743
Hidden neuron 2 (k = 2)	11.4264	−18.1075	−0.9497	−0.1099	20.8869	
Hidden neuron 3 (k = 3)	24.9425	15.2804	13.5236	0.1446	38838	
Hidden neuron 4 (k = 4)	26.6906	1.1618	−14.609	0.2608	10.2778	
Hidden neuron 5 (k = 5)	0.5598	2.1791	−0.8329	−0.5202	−1.8638	
Hidden neuron 6 (k = 6)	1.131	0.7402	−0.4105	−0.6329	0.9429	

Table 4. Cross-correlation of input and output for reduction factor

	B/L	*e/B*	*D_f/B*	RF_{expt}
(B/L)	1	−0.1	0	0.012
(e/B)		1	0	0.975
(D_f/B)			1	0.167
RF_{expt}				1

Table 5. Relative importance of different inputs as per Garson's algorithm and connection weight approach

Parameters	Garson's algorithm		Connection weight approach	
	Relative importance	Ranking of input as per relative importance	S_i values as per connection weight approach	Ranking of input as per relative importance
B/L	36.41	2	8.6	1
e/B	45.08	1	−1.06	3
D_f/B	18.51	3	1.38	2

7 Conclusion

Based on developed neural network model, the following conclusions may be drawn.

1. The errors are distributed evenly along the centerline as per residual analysis. It can be concluded that the network was well trained and can predict the reduction factor *RF* with reasonable accuracy.
2. Based on Pearson correlation coefficient, it was observed that e/B is the most important input parameter followed by B/L and D_f/B and as per the Garson's algorithm e/B is the most important input parameter followed by B/L and D_f/B.
3. The developed ANN model could explain the physical effect of inputs on the output, as described in NID. It has been observed that e/B is inversely related to *RF*, whereas B/L and D_f/B are directly related to *RF*.
4. A model equation is developed based on the trained weights of ANN.

Refernces

Behera, R.N., et al.: Prediction of ultimate bearing capacity of eccentrically inclined loaded strip footing by ANN part 1. Int. J. Geotech. Eng. (2013). doi:10.1179/1938636212Z.00000000012

Garson, G.D.: Interpreting neural-network connection weights. Artif. Intell. Expert **6**(7), 47–51 (1991)

Goh, A.T.C.: Seismic liquefaction potential assessed by neural network. J. Geotech. Eng. **120**(9), 1467–1480 (1994)

Goh, A.T.C., et al.: Bayesian neural network analysis of undrained side resistance of drilled shafts. J. Geotech. Geoenvironmental Eng. **131**(1), 84 (2005). doi:10.1061/(ASCE)1090-0241

Guyon, I., Elisseeff, A.: An introduction to variable and feature selection. J. Mach. Learn. Res. **3**, 1157–1182 (2003)

Hornik, K., et al.: Multilayer feed forward networks are universal approximators. Neural Netw. **2**, 359–366 (1989)

Olden, J.D., et al.: An accurate comparison of methods for quantifying variable importance in artificial neural networks using simulated data. Ecol. Model. (2004). doi:10.1016/j.ecolmodel.2004.03.0.013

Ozesmi, S.L., Ozesmi, U.: An artificial neural network approach to spatial modeling with inter specific interactions. Ecol. Model. (1999). doi:10.1016/S0304-3800(98)00149-5

Shahin, M.A., et al.: Predicting settlement of shallow foundations using neural network. J. Geotech. Geoenvironmental Eng. **128**(9), 785 (2002). doi:10.1061/(ASCE)1090-0241

Wilby, R.L., et al.: Detection of conceptual model rainfall—runoff processes inside as artificial neural network. Hydro. Sci. **48**(2), 163–181 (2003)

Author Index

S.K. Shukla and E. Guler (eds.), *Advances in Reinforced Soil Structures*,
Sustainable Civil Infrastructures, DOI 10.1007/978-3-319-63570-5

MIX
Papier aus verantwortungsvollen Quellen
Paper from responsible sources
FSC® C105338

If you have any concerns about our products,
you can contact us on
ProductSafety@springernature.com

In case Publisher is established outside the EU,
the EU authorized representative is:
Springer Nature Customer Service Center GmbH
Europaplatz 3, 69115 Heidelberg, Germany

Printed by Libri Plureos GmbH
in Hamburg, Germany